Acoustics for Engineers

J. D. Turner and A. J. Pretlove

MACMILLAN

First edition 1991

Published by
MACMILLAN EDUCATION LTD
Houndmills, Basingstoke, Hampshire RG21 2XS
and London
Companies and representatives
throughout the world

Printed in Hong Kong

British Library Cataloguing in Publication Data
Turner, J. D. (John David), *1955*–
Acoustics for engineers.
1. Acoustic engineering
I. Title II. Pretlove, A. J.
620.2
ISBN 0–333–52142–0
ISBN 0–333–52143–9 pbk

To our children,
Jack,
Gareth,
John,
Steve,
and James,

who, very early in their lives,
taught us the importance of noise control.

Contents

Preface

Sound is a phenomenon which all human beings live with from birth, and perhaps even before. It is a familiar medium of communication with our fellow man and it includes speech and the pleasures of music. Hearing is one of our primary senses, at once vitally valuable to our development and of infinitely varied interest. Sound can also be an irritating nuisance. All of this is taken for granted by most people. Because it is so familiar, relatively few will question how it works. Indeed, the basic physics of sound was not understood at all two hundred years ago. The pioneering work of early physicists such as Helmholtz (1821–1894) and Lord Rayleigh (1842–1919) defined sound as a form of wave motion and the wave theory explained various well-known phenomena such as diffraction (hearing a sound round a corner) and echoes (reflected sound with time delay). Once sound was recognised as a form of wave motion it could be brought into the family of wave phenomena and analogies could then be made with light, electromagnetic radiation and so forth, to the mutual benefit of all. An understanding of the physics of sound is particularly essential to the practising engineer whose task is often to design noise control devices for the factory, the office and at home.

Noise can be just as damaging as any other form of pollution. It may cause irritation or annoyance, and can be a danger to health. For these reasons, legal limits on the exposure to and emission of noise exist in most countries. As well as these health and safety aspects of noise, our acoustic environment is under constant threat from the increase in use and power of vehicles and machinery of all sorts. The benefits of a quiet environment, which was once accepted as the norm, are increasingly appreciated. The preference therefore is for quieter products whenever possible, and this is often reflected by the selling advantage the quiet product has over its noisy competitor.

Noise measurement is an important weapon in the armoury of the engineer concerned with diagnosing or monitoring the condition of machinery. The wheel-tapper who used to be employed on railways is a familiar example. We all know that faulty machines often emit unusual sounds prior to failure. The sounds made by machinery in normal use can be monitored to give advance warning of maintenance requirements. Measurement is therefore a subject of fundamental importance to the noise control engineer. Without it, there is no means of quantifying the extent of a noise problem. A modern measurement system is considerably different from its counterpart of even ten years ago. The only component which is the same is the microphone itself. The rest of the system has, with the advent of the ubiquitous microprocessor or microcomputer, gone digital. In acoustical analysis, digital signal processing is now of paramount importance because of its power, accuracy and speed. For this reason, a considerable proportion of this book is devoted to the fundamental principles of measurement and analysis. This includes not only Fourier analysis but also other useful aspects of signal processing.

Noise costs money. Processes which are inevitably noisy will require expenditure on noise insulation. Excessive noise levels can lead to annoyance or to a loss of concentration, and productivity may fall. The control of noise is therefore important not only to prevent hearing damage and maintain a pleasant environment, but also from the economic point of view. A novel strategy for tackling noise control problems is given in the last chapter of this book. The task of diagnosing and correcting noise problems in the most effective manner is one which, in the past, has received little attention in a formal way. It is hoped that this will provide the reader with an accurate method for defining the best medicine to be prescribed for the noisy patient.

The book is therefore unusual in integrating modern measurement techniques with fundamentals of applied acoustics and the determination of the most effective treatment of noise problems. By this means an up-to-date statement of the basics of noise and noise control is presented. The book aims to describe the basic theory and its practical applications in an easily digestible way with plenty of examples drawn from a wide variety of backgrounds.

Apart from being of general interest, the book is intended to be of value to students on courses of higher education and perhaps particularly to student engineers who, in their later professional lives, will often be faced with problems associated with noise. Many engineering degree courses now recognise the importance of noise control to the practising engineer and this book is designed to fulfil a need for a text which covers, in reasonably concise and up-to-date form, the major aspects of the subject.

The authors acknowledge the constructive comment and informed criticism of their many interested colleagues at Reading, Southampton and elsewhere in the preparation of this book. Particular thanks are due to Martyn Hill, who patiently read the manuscript and made many helpful comments.

Acknowledgements

The authors wish to thank Professors M. T. Thew of Southampton University and A. G. Atkins of Reading University, for their encouragement of the work on which this book is based. We would also like to acknowledge the assistance provided by Mr M. Hill and Professor F. Fahy.

The index was produced with the aid of an extremely efficient piece of software called MACREX. The authors were lent a copy of MACREX by its creators, and would like to record their gratitude.

Permission for the reproduction of a number of figures and tables is acknowledged from Bruel and Kjaer Ltd. These are indicated in the text.

Every effort has been made to trace all copyright holders but if any have been inadvertently overlooked the publishers will be pleased to make the necessary arrangement at the first opportunity.

1. Basic concepts of noise and vibration

Take care of the sense, and sounds will take care of themselves.
Lewis Carroll, *Alice in Wonderland*

Noise is a problem in many parts of industry, partly because of its direct effects on the workforce and partly because of its impact on the local community. Acoustics is important in design because many machines have to meet noise and vibration specifications. It is good engineering practice to design quiet machinery since, all other things being equal, the quiet machine is preferred by the customer to the noisy one. Noise is often an indicator of damaging vibration within a machine.

Noise at work causes hearing impairment, and legal limits are imposed because of this. However, there are also less obvious undesirable effects. Noise can cause irritation, which may lead to irrational behaviour with consequent loss of safety, or even hearing damage. It can cause loss of concentration which is likely to impair safety and productivity.

It is important therefore to be able to assess (measure) noise in relation to its damaging effects, to diagnose sources of noise and to be able to take remedial action. This requires a knowledge of the basic physics of noise generation and radiation and human reaction to noise. These topics are what this book is about. It is a basic 'tool kit' for the understanding of noise problems, and is intended to provide 'literacy' in acoustics, measurement and noise control. The subject is, in fact, a very broad one and includes many advanced and difficult topics for the engineer, such as the problem of silencing aircraft engines. Some of the advanced topics are beyond the scope of this book. Many significant advances in noise control have been made in recent times, and the subject is very active on the current engineering research agenda.

In this opening chapter the general nature of sound and vibration and the derivation of the fundamental laws of sound transmission are described. The acoustical analysis is for one-dimensional waves, which can be thought of as sound travelling in a tube. A brief introduction to the use

1

of analogies is given by comparing the radiation of sound energy in a tube to the passage of electric current in a simple resistive electric circuit. The conventional decibel scale is introduced and the simplest ideas of accounting for human reaction to noise are described. This is followed in chapter 2 by a more detailed account of the human response to noise.

At the end of the chapter the analysis of vibration for a simple system is given together with its application in the vibration of beams and plates. Beams are often responsible for the transmission of vibration energy which subsequently is converted into sound, for example, as in steel and concrete frame buildings. Vibrating plates can radiate sound directly into the surrounding air.

1.1   THE GENERAL NATURE OF SOUND

Sound is a wave motion which is in some ways similar to the moving ripples that appear on the surface of a pond when a stone is thrown in. A sound wave is characterised by pressure disturbances superimposed on atmospheric pressure, and by accompanying oscillations of the air particles. The pressure disturbances at a point act in all directions. Sound pressure is therefore sometimes described as a scalar quantity by those familiar with vector analysis, or as a hydrostatic pressure by those familiar with fluid mechanics. On the other hand, the particle velocity is a vector quantity which usually lies along the line from the source to the point in question. The disturbances travel out from the source, but the medium which transmits them (air) does not travel and simply oscillates about a fixed point. Another way of viewing this is to imagine each oscillating particle nudging its neighbour and thus passing on the message. The rate at which the message travels is the speed of sound (c), which for air at 20°C is about 340 m/s. The wave travels because the medium is both massive and elastic and the speed of sound depends crucially on these two qualities.

It is important to understand that the pressure disturbance involves the outward transmission of energy from a sound source, though the amounts of energy involved are usually *very* small. For this reason not much effort is required to speak (actually less than one-thousandth of a watt) and it is impossible to boil a kettle of water by shouting at it. In three dimensions the pressure disturbance will diminish as the wave travels outwards from a source, since the initial finite amount of energy is gradually spreading itself over a larger and larger area. The radiation of sound in three dimensions is described in detail in chapter 3. If the wave is confined to one dimension, for instance in a tube of fluid or gas or in a solid rod, then the amplitude of the pressure disturbance does not diminish greatly since the only energy loss mechanism operating is the friction between the molecules of the medium and the tube. This is why the 'speaking tubes' which were once used on ships and in some large buildings were so successful.

The fact that sound is a form of wave motion implies that these waves, like light waves, are subject to reflection, refraction and diffraction. It is important to have a clear idea of the differences between these phenomena. For sound in air, *reflection* occurs when the acoustic wave impinges on a boundary between the air and a different material, such as a concrete wall and also in a pipe when there is a change of section. It occurs for much the same reason as causes a ball to bounce on the ground: there is a sudden change in *impedance* at the boundary such that the wave energy is largely turned back whence it came. This is dealt with in detail in chapter 7. *Refraction*, on the other hand, means a bending or deflection of the wave away from a straight path of propagation. Refraction is always caused by changes in the absolute wave velocity. In acoustics these are usually gradual changes caused, for example, by temperature or wind gradients. The consequences of these on sound in the open air are described at the end of chapter 3. *Diffraction* is quite distinct from refraction and may be described as the gradual spreading out of an acoustic wave in an angular sense. So, for example, when sound travelling in a narrow parallel-sided beam in a uniform round pipe reaches an open end and emerges into an open space it will not continue as a parallel-sided beam but will become a divergent beam (there will also be a measure of reflection back up the pipe). The amount of divergence in the emerging beam will depend on the ratio of the acoustic wavelength to the pipe diameter. Very short wavelength sounds will exhibit very small divergence and vice versa. Diffraction is very important in the design of noise screens or barriers, as is described towards the end of chapter 3.

In the analysis of sound 'harmonic' waves are often used, that is, pressures which vary sinusoidally in both time and space. The variation in time leads to the concept of frequency (f in Hz or cycles per second) whilst the variation in space is defined by the wavelength (λ). The two are connected by the speed of sound. Thus

$$c = f\lambda \tag{1.1}$$

For example, on the musical scale, middle C is 256 Hz so the wavelength will be about 1.34 m. The wavelength is, in fact, the most important characteristic of a sound when it comes to making calculations concerned with radiation, as will be seen. Sounds in real life are very rarely harmonic. The sounds from a musical instrument usually consist of a base note (or fundamental) upon which is superimposed a series of higher harmonics having frequencies which are multiples of the base frequency. Noise from some forms of machinery, such as gearboxes for example, is often of a similar nature though usually much richer in higher harmonics, and it includes sounds without a harmonic relationship. Other sound sources, such as gas burners and jet engines, produce a total sound which contains a random and variable mixture of many frequencies. Such sounds are called 'noise'. If there is a uniform spread of sound energy over all frequencies, such noise is called 'white noise', by analogy with 'white light'. Thus it is seen that the character of a noise is determined to a large extent by the distribution of its energy as a function of frequency. A graph of this is called the noise spectrum, and an example is shown in figure 1.1

For all sounds other than those of extremely high intensity, the acoustic waveform is preserved as the wave spreads out. It is therefore convenient to consider sounds as a superposition of harmonic components of different frequencies. An analysis of a measured sound in this manner can be made by simple frequency filtering in narrow or broad bands. Broad octave bands are often used in simple measurements. Much more accurately a mathematical Fourier Analysis can be made which will rely on digital measurement systems and appropriate software in a computer. Such analyses as these are very useful in diagnostic work and the results are presented in the form of an energy spectrum of the noise, see figure 1.1. The octave band spectrum is found by averaging the narrow band spectrum over intervals of one octave. An octave interval spans from a lower frequency up to twice that frequency and the nomenclature is derived from the musical scale. For example, the 500 Hz octave band starts at 353.6 Hz and extends up to 707.2 Hz.

It is perhaps worth stating that the narrow band spectrum for a harmonic sound consists of a single vertical line at the relevant frequency. The energy spectrum is very characteristic of a noise and very distinctive. The human ear can make these sort of distinctions with great sensitivity: it can very easily differentiate between a violin and a cello playing the same note. At an even more sensitive level a speaker can often be identified (even on the telephone, with all its distortions) from the utterance of one or two words.

Spectra such as these form one of the most powerful tools available to the acoustic engineer. They can be thought of as a 'fingerprint' by which the source of a noise can often be identified. The narrow band spectrum may be essential for problem diagnosis because of the great amount of detail which it displays. By contrast, the octave band spectrum is completely defined by a handful of numbers, is quickly measured and often provides an adequate description of the noise.

1.1.1 Derivation of the one-dimensional wave equation

The one-dimensional analysis which follows is essentially for sound travelling in a tube. However, it is more straightforward for it to be derived

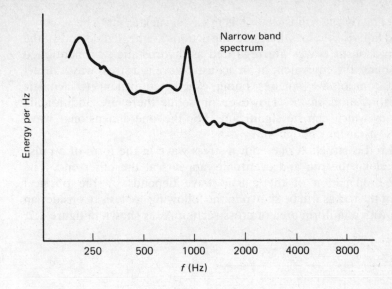

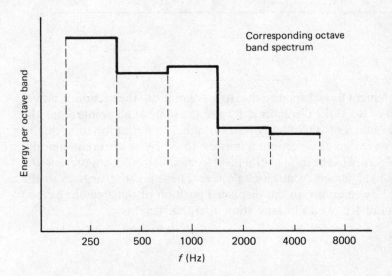

Figure 1.1 Narrow and broad band spectral representations

initially in terms of stress waves in a solid bar. The results are then related to a sound wave in air. One-dimensional sound occurs in a speaking tube or in an organ pipe, for example. A sound wave in three-dimensional space, far from its source, may also be regarded as one-dimensional since the wavefront is approximately flat. Sound waves may pass through any medium, in liquids and solids as well as in gases, such as air. The viscosity of gases and liquids is so low that shearing actions are negligible and so the pressures in acoustic waves are regarded as 'hydrostatic', as mentioned earlier. In solids, the equivalent of an acoustic wave is a stress wave, direct tensile and compressive stresses being exactly equivalent to acoustic pressure disturbances in air. However, in solids there are additionally shear stresses which can be significant. For the one-dimensional wave studied here shear may be ignored.

If a solid rod is struck at one end, a stress wave in the form of a pulse propagates along the rod and eventually appears at the other end. The velocity of propagation of the elastic wave depends on the physical properties of the rod as will be shown in the following analysis. Consider an infinite rod with a uniform area of cross-section A, as shown in figure 1.2.

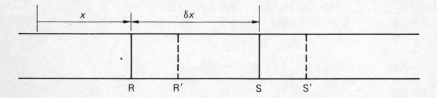

Figure 1.2

When the element δx is displaced due to wave motion, the section R moves a distance u to R'. If the displacement were the same at all points along the rod it would undergo whole-body motion rather than deformation due to the stress wave, so the section S moves to S' by a different amount $(u + \delta u)$. Because of the differential displacement, stresses are present and these can be calculated from Hooke's law. These are the stresses in the stress wave. By reference to the displaced position of the element δx, see R' S' on figure 1.2, we can easily show that the strain is

$$\epsilon = \frac{\partial u}{\partial x}$$

at the left-hand end of the element and hence the stress is

$$\sigma = E \frac{\partial u}{\partial x}$$

(1.2)

If the limits are properly drawn this will be the stress at R'. However, the stress is a variable quantity and will be different at the other end of the element. The value there will be

$$\sigma + \frac{\partial \sigma}{\partial x} \delta x$$

There is thus no static equilibrium of the element and acceleration will occur. Reference to the free-body diagram for the element, figure 1.3, permits the equation of dynamic equilibrium to be defined.

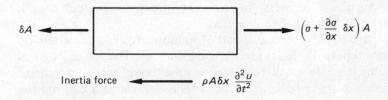

Figure 1.3

Hence

$$A\delta x \frac{\partial \sigma}{\partial x} - \rho A\delta x \frac{\partial^2 u}{\partial t^2} = 0$$

or

$$\frac{\partial \sigma}{\partial x} - \rho \frac{\partial^2 u}{\partial t^2} = 0$$

but from equation 1.2

$$\sigma = E\partial u/\partial x$$

so

$$E \frac{\partial^2 u}{\partial x^2} - \rho \frac{\partial^2 u}{\partial t^2} = 0 \qquad (1.3)$$

This is the classical one-dimensional wave equation which is usually expressed as

$$\frac{\partial^2 u}{\partial t^2} - c^2 \frac{\partial^2 u}{\partial x^2} = 0 \qquad (1.4)$$

in which c is the speed of wave propagation. By comparison we have:

$$c = \sqrt{(E/\rho)} \qquad (1.5)$$

The well-known general solution to equation 1.4 is

$$u = f(x - ct) + g(x + ct) \tag{1.6}$$

The general function $f(x - ct)$ represents a positive-going wave of arbitrary but *unchanging* shape travelling at speed c. This is why when we shout 'hello' to our friend across the street, he receives the waveform in unchanged shape and registers 'hello' rather than some other sound. Similarly $g(x + ct)$ is a negative-going wave.

It is instructive to calculate the speed of sound-waves in a few common engineering materials. Table 1.1 gives some examples. The velocity figures for solids in the table do not quite agree with the 'book' values, which are for compression waves in bulk materials. This is because in a thin rod the effect of Poisson contraction reduces the stiffness of the material.

Table 1.1   Acoustic properties for a selection of common materials.

Material	Young's Modulus	Density	Wave velocity	Characteristic impedance
	(N/m^2)	(kg/m^3)	(m/s)	(Ns/m^3)
Nickel	2.2×10^{11}	8800	4970	4.4×10^7
Steel	2.1×10^{11}	7800	5200	4.1×10^7
Aluminium	6.9×10^{10}	2720	5030	1.4×10^7
Glass	6×10^{10}	2400	5000	1.2×10^7
Concrete	3×10^{10}	2400	3500	8.4×10^6
Lead	1.7×10^{10}	11400	1230	1.4×10^7
Hardwood	1×10^{10}	600	4000	2.4×10^6
Nylon	2×10^9	1140	1320	1.5×10^6
Water	2.3×10^9	1000	1500	1.5×10^6
Mineral oil	1.6×10^9	800	1400	1.1×10^6
Air	1.4×10^5	1.2	340	407

1.1.2   Sound in air

The analysis above is fine for a solid bar but it is not obvious how it applies to air because there is the problem of defining the Young's Modulus, E. There is also a minor difference of sign convention. In a solid it is usual to regard *tensile* stresses as positive whereas acoustic *compressive* pressures are taken to be positive.

In the passage of a sound wave compression of the air is so rapid that the adiabatic gas law must apply, that is:

$$PV^\gamma = C \text{ (a constant)} \tag{1.7}$$

In the acoustic wave there are perturbations of both pressure and volume about fixed static values and we may express this as:

$$P = P_0 + \delta P$$
$$V = V_0 - \delta V \text{ (taking the sign convention into account)}$$

In these equations P_0 is the atmospheric pressure and δP is the acoustic perturbation corresponding to the stress σ in the analysis for the solid bar. Differentiation of the adiabatic equation gives:

$$\delta P V_0^\gamma - P_0 \gamma V_0^{\gamma-1} \delta V = 0$$

or

$$\delta P = P_0 \gamma \left(\frac{\delta V}{V_0} \right)$$

This is the equivalent of Hooke's Law for the air and it can be seen that the equivalent Young's Modulus is

$$E_{\text{air}} = P_0 \gamma \tag{1.8}$$

It follows, by analogy with the analysis for the solid bar, that

$$c = \sqrt{\left(\frac{P_0 \gamma}{\rho} \right)} \text{ or } c = \sqrt{(\gamma R T)} \tag{1.9}$$

where R is the gas constant. The principal conclusion to be drawn from this is that c depends only on the absolute temperature, T. It also follows that the elastic modulus can now be expressed in a slightly different way, for:

$$P_0 \gamma = \rho c^2 \tag{1.10}$$

The acoustic pressure, up to this point denoted by δP, will henceforth be described by the symbol p.

Using the above formulation for the effective modulus and the strain value from the solid rod analysis:

$$p = -\rho c \frac{\partial u}{\partial x} \tag{1.11}$$

The minus sign is necessary because of the acoustic sign convention. The speed of sound in air can be calculated as follows, and is much less than the value for a solid like steel:

$$P_0 = 10^5 \text{ N/m}^2 \qquad (1.12)$$
$$\gamma = 1.4$$
$$\rho = 1.18 \text{ kg/m}^3$$

$$c = \sqrt{\left(\frac{P_0\gamma}{\rho}\right)} = 344 \text{ m/s} \qquad (1.13)$$

Apart from the formulae for the speed of sound the important conclusions from the foregoing analysis are:

- The speed of sound is the same for all waves of whatever wavelength. Thus, distortion due to dispersion does not occur. In simpler language, if we listen to speech through a long tube, the components of the sound of various frequencies will all travel at the same speed and the speech will not be distorted. The same is true for light waves, electromagnetic waves and waves in a stretched string, but it is untrue for waves on the beach or bending waves in beams, both of which suffer distortion as a result of dispersion.

- Sound in a given gas (for which R is constant), such as the atmosphere, travels at a speed solely dependent on the square root of the absolute temperature (equation 1.9). This means that sounds are refracted or bent in travelling from gas at one temperature to gas at another, just as light is bent in going from one medium to another. Under the meteorological condition known as a temperature inversion, acoustic mirages may form as a result of this bending. On the other hand, under normal atmospheric conditions with temperature decreasing with height, sound rays are bent upwards thus creating a shadow zone near the ground. Wind and wind gradients will also have an important influence on the concentration and dilution of sound in the open air.

1.1.3 Relationship between pressure and velocity: the energy in a sound wave

It is obvious that energy has to be expended in order to propagate a sound wave and, in fact, energy is radiated away from the source at the speed of sound. The sound wave carries energy with it. To calculate this energy we need to look again at the mechanics of what happens to the one-dimensional sound wave travelling along in the 'speaking tube'.

Referring to figure 1.4, the work done at the boundary x by the left-hand mass pushing against the right-hand mass is $pA\delta u$ in a small time δt. So, the *rate* of doing work is

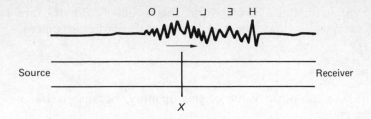

Figure 1.4

$$pA \frac{\partial u}{\partial t}$$

The acoustic pressure, p, and the particle velocity in this expression are directly related. For the positive-going wave shown in figure 1.4, displacement u is expressed:

$$u = f(x - ct)$$

Function f is the waveform representing the word 'hello' shown in figure 1.4. The particle velocity is thus

$$\frac{\partial u}{\partial t} = -cf'(x-ct)$$

The pressure is obtained from:

$$p = -\rho c^2 \frac{\partial u}{\partial x} = -\rho c^2 f'(x-ct)$$

Hence

$$p = \rho c \frac{\partial u}{\partial t} \tag{1.14}$$

This very important relationship shows that the pressure and velocity are directly proportional to one another for any waveform at any instant. For a negative-going wave

$$p = -\rho c \frac{\partial u}{\partial t} \tag{1.15}$$

Now let us return to the rate of doing work. This is equivalent to the rate of radiation of energy along the tube and extending the earlier argument:

$$pA \frac{\partial u}{\partial t} = \frac{Ap^2}{\rho c}$$

It is sensible to take the mean value of this quantity, because it varies rapidly, and this gives:

$$\text{Mean rate of doing work} = \frac{A\overline{p^2}}{\rho c}$$

with

$$\overline{p^2} = \text{mean square pressure}$$

The quantity $\overline{p^2}/\rho c$ is called the INTENSITY and is measured in W/m^2. It is the time-averaged rate of radiation of energy per unit area. The symbol for intensity is I, thus

$$I = \frac{\overline{p^2}}{\rho c} \tag{1.16}$$

1.1.4 Electric circuit analogy

We have seen that acoustic energy is being transferred from a source to the left of x to a receiver to the right of x by means of radiation in the form of acoustic waves. A useful analogue can be made between this acoustic system and the simple electric circuit shown in figure 1.5

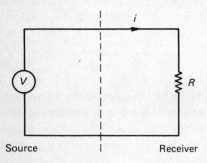

Source   Receiver

Figure 1.5

The voltage V is taken to be analogous to the acoustic pressure; the electric current i is analogous to the acoustic particle velocity, thus

$V \equiv$ pressure p
$i \equiv$ particle velocity $\partial u/\partial t$

The analogy is being applied here to a unit cross-sectional area of the pipe. Equation 1.14 is the acoustic equivalent of Ohm's Law:

$$p = \rho c \, \frac{\partial u}{\partial t} \text{ is equivalent to } V = Ri$$

Hence, the resistance R is analogous to ρc, a property of air (or of any other material).

In general (that is, not just for one-dimensional waves) the constant of proportionality between pressure and particle velocity is called the 'acoustic impedance' (symbol Z) and may be a complex quantity as it would be for electric circuits containing capacitance and inductance. The special quantity ρc, which is valid only for one-dimensional or plane fronted-waves, is called the specific acoustic impedance or the characteristic impedance. The analogy itself also leads directly to a value for power transfer bearing in mind that it is valid for a UNIT AREA of the cross-section of our pipe. For, in the electrical analogy

$$\textbf{Power} = \overline{i^2}R = \overline{V^2}/R$$

and this is analogous to $\overline{p^2}/\rho c$ which is the INTENSITY as defined before.

Another important interpretation or use of the concept of impedance is concerned with the radiation from and the forces on mechanical drivers such as, for example, loudspeakers or panelling on a machine tool. Referring back to figure 1.4, the left-hand section of air (source) up to the boundary at x can be replaced by a mechanical piston (like a loudspeaker) performing exactly the same motion. The right-hand section will be unaware of this deception and will continue to carry the radiated word 'hello' as and when it arrives. The relationship between force and velocity at the surface x will be exactly as before and the mechanical driver piston will therefore have to be activated by sufficient force to produce the required velocities. The ρc quantity is then seen as a 'driving point impedance' which makes demands on the mechanical capacities of the piston. Conversely, if it is known that the vibrational velocities of a surface have given values, the corresponding pressures can be calculated and, therefrom, the amount of acoustic energy being radiated. Calculations of this kind are often made by acoustic engineers based on measurements of vibrational velocity of machine surfaces. An example is given later in this chapter.

As can be shown by use of the electric analogy, optimum energy transfer takes place when the mechanical impedance of the source matches the specific impedance of the air. In practice, the impedance of mechanical sources is usually several orders of magnitude greater than ρc so that the radiation process is mercifully inefficient. In the design of loudspeakers, on the other hand, acoustic efficiency is to be achieved, so great effort goes into keeping the mechanical impedance as low as possible and this generally means using light cones made of paper. Even so, the electro-acoustic efficiency of loudspeakers is rarely better than 1 per cent and usually lower than 0.1 per cent.

The value of the characteristic impedance for air is

$$\rho c = 407 \text{ Ns/m}^3 \tag{1.17}$$

Table 1.1 lists values for the characteristic impedance for the range of materials shown there.

1.1.5   The decibel scale

The human ear, which is in some ways like a microphone, does not respond linearly to sound. It responds on a logarithmic scale. That is to say, a repeated doubling of sound intensity is perceived approximately as a repeated addition of a constant amount. This fact, together with the very large range of audible sound intensities, leads to the use of a logarithmic scale for sounds called the decibel scale. At the lower end of the scale, the ear can just detect a sound in which the particle displacement is of the order of molecular size. This indicates the remarkable sensitivity of the human ear. At the other end of the scale, the pain threshold is about 10^{14} times greater in acoustic intensity (or about 10^7 times greater in pressure or velocity).

Expressed in decibels (dB) the intensity I is:

$$\textbf{Intensity (dB)} = 10 \log_{10} \left(\frac{I}{I_0} \right) \tag{1.18}$$

where I_0 is the intensity of the average lower threshold of audibility. From equation 1.16 this can be expressed as

$$\textbf{Intensity in dB} = 10 \log_{10}(\overline{p^2}/\overline{p_0^2}) = 20 \log_{10}(\overline{p}/\overline{p_0}) \tag{1.19}$$

$\overline{p_0}$ here is the acoustic pressure at the average lower threshold of audibility and by convention is taken to be

$$\overline{p_0} = 2 \times 10^{-5} \text{ N/m}^2 \tag{1.20}$$

From this, using equations 1.16 and 1.17, the threshold intensity may be calculated:

$$I_0 = 10^{-12} \text{W/m}^2 \tag{1.21}$$

This is a remarkably low power transfer for the ear to detect. It represents a power of about 10^{-16} Watts entering each ear. It is of the same order as the electromagnetic field intensity at a radio receiver from a local station and this usually requires some amplification before becoming audible.

For the decibel scale, as defined above, it is approximately true that:
(a) a doubling of intensity ($\sqrt{2} \times$ pressure) gives an increase of 3 dB;
(b) a tenfold increase in intensity gives an increase of 10 dB.

The smallest change which the ear can detect is between 1 and 3 dB depending on the frequency and the loudness level. The maximum possible noise in a normal atmosphere is limited by the fact that on negative excursions the sound wave pressure may not go below a vacuum.

Table 1.2

SPL (dB)	Comment or typical example	I (W/m^2)
194	Maximum possible without going below a vacuum	10^8
140	Gunfire at the gunners ear; jet engine at 30 m	10^2
130	Threshold of pain	10
120	Pneumatic breaker at 1 m; Concorde take-off at 500 m	1
110	Rock concert; disco; car horn at 1 m	0.1
100	Discomfort; hearing damage long term; heavy industry	10^{-2}
90	Inside an underground train	10^{-3}
80	Inside a bus; shouting at 1 m	10^{-4}
70	Heavy city traffic at the kerbside	10^{-5}
60	Busy restaurant or department store; conversation	10^{-6}
50	Quiet car interior; average business office	10^{-7}
40	Quiet residential neighbourhood; living room	10^{-8}
30	Library; empty lecture theatre	10^{-9}
20	Recording studio; bedroom at night; whisper at 1 m	10^{-10}
10	It was so quiet you could hear a pin drop (just)	10^{-11}
0	Threshold of hearing (average for a young person)	10^{-12}

The decibel as the absolute physical measurement of pressure is often referred to as the sound pressure level (SPL, also referred to as L_p) in order to differentiate it from some of the subjective scales which will be discussed shortly, which are often referred to (incorrectly) as just dB. Table 1.2 shows the audible range of *SPL*-values together with a corresponding description of typical sounds at these levels and the intensity values.

Example: A sound power calculation from mechanical measurements

A small flat panel of area 0.25 m^2 fills the end of a closed uniform duct and it vibrates sinusoidally at 200 Hz. The peak value of the surface acceleration amplitude is measured to be 15.8 m/s^2 using a miniature accelerometer. Estimate the radiated sound power from the panel and the *SPL* in the duct.

Solution

The motion of the panel is of the form

$$u = B\cos\omega t$$

so the acceleration amplitude is

$$|a| = B\omega^2$$

However, to calculate intensity the velocity amplitude is required and this is

$$|v| = B\omega$$

so

$$v = |a|/\omega$$

At 200 Hz

$$\omega = 2\pi \times 200 = 1257 \text{ rad/s}$$

Hence

$$|v| = 15.84/1257 = 0.0126 \text{ m/s}$$

and the r.m.s. velocity is

$$v = 0.0126/\sqrt{2} \text{ m/s}$$

Intensity at the surface

$$I = \overline{p^2}/\rho c = \overline{v^2}\rho c$$

$$= \frac{0.0126^2}{2} \times 407$$

$$= 0.0323 \text{ W/m}^2$$

Hence the radiated sound power is

Power $= 0.0323 \times 0.25 = 0.008$ W

The intensity in the duct is assumed to be the same as at the panel surface. Thus the *SPL* in dB is

$$10 \log_{10} \left(\frac{0.0323}{10^{-12}} \right) = 105.1 \text{ dB}$$

1.1.6 Adding decibels

If there are two or more sources of noise, and it is known how much noise each makes individually, then there is a need for some rule for adding their effects. Provided that the sources are not 'correlated' it is fairly obvious that the energies will be additive. If noises arise basically from a common source they may well be correlated and the energies will not then add directly for

$$p_t = p_1 + p_2 + \ldots \text{ from each source}$$

$$\overline{p_t^2} = \overline{p_1^2} + \overline{p_2^2} + \ldots + \overline{2p_1p_2} + \ldots$$

The time average products, such as $\overline{p_1}\ \overline{p_2}$, are zero if the sources are uncorrelated, but not otherwise.

Thus the required method is to calculate the intensity for each source, add, and then recalculate the total dB. This is a rather complicated procedure which is made easy by using a computer. In practice, it can also be made simple by the use of the special adding nomogram, figure 1.6, which can be used in hand calculations. The noise energies shown in the bands of a noise spectrum can be added in the same way to give the total sound pressure level (*SPL*).

Difference between two noise levels in decibels

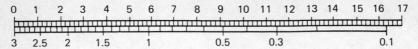

Addition (dB) to be made to higher of two noise levels to obtain combined noise level (add further noise levels in turn to progressive total)

Figure 1.6

Example: Use of the nomogram, figure 1.6

The noise from a machine is measured in octave bands, the values being given in the table below. As a check, an overall *SPL* measurement is made and has a value of 78.5 dB. Use the nomogram to check that the sum of the octave band values gives the same overall *SPL*.

Octave band centre frequency (Hz)	Banded SPL-value (dB)
125	75
250	72
500	65
1000	71
2000	66
4000	63
8000	55

Solution

The process is started by adding 75 dB and 72 dB, the top two entries in the table. The difference between the values is 3 dB. This is looked up along the top scale of the nomogram. The corresponding addition to be made (caution: the scale moves in the opposite sense) is 1.8 dB to the nearest 0.1 dB. Hence the sum is 76.8 dB. The calculation can proceed from left to right, adding pairs of values, as follows:

75 ⎫
      ⎬ 76.8 ⎫
72 ⎭   ⎪
             ⎬   78.0 ⎫
65 ⎫   ⎪   ⎪
      ⎬ 72.0 ⎭   ⎪
71 ⎭   ⎬   78.4
                              ⎪
66 ⎫   ⎪
      ⎬ 67.8 ⎫   ⎪
63 ⎭   ⎪   ⎪
             ⎬   68.0 ⎭
55 ⎫   ⎪
      ⎬ 55 ⎭
55 ⎭

This result is close enough to the measured value to be a satisfactory check. In the addition process the same result should be achieved regardless of the way in which the initial pairing is done.

1.2   BASIC VIBRATION THEORY FOR SINGLE DEGREE OF FREEDOM SYSTEMS

1.2.1   Free vibration

A single degree of freedom (abbreviated to SDOF) linear system is described by the following quantities:

x = displacement
m = mass
c = viscous damping coefficient
k = stiffness

It is modelled as in figure 1.7 but may look very different in practice. For example, it may be a model for a cantilever beam or the wall of a room. Free vibration of such a system is characterised by the homogeneous differential equation of motion:

$$m\ddot{x} + c\dot{x} + kx = 0 \qquad (1.22)$$

The solution to this equation is a sinusoidal vibration, the frequency of which is strongly dependent on k and m. The value of c affects the decay of

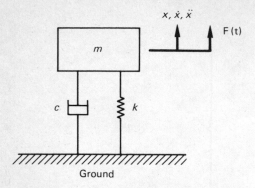

Figure 1.7

the vibration but has a relatively weak influence on the frequency of vibration, negligible in most cases of structural vibration. The damping in real structures is often not strictly viscous (linear) but in most such cases an equivalent viscous damping coefficient can be used with satisfactory results. If $c = 0$ the *circular natural frequency* is

$$\omega_1 = \sqrt{\left(\frac{k}{m}\right)} \ \text{rads/s} \tag{1.23}$$

Circular frequencies are related to cyclic frequencies (f in Hz) by the relationship

$$\omega = 2\pi f \tag{1.24}$$

If the damping coefficient $c > 0$ then the non-dimensional *damping ratio* is defined as

$$\zeta = \frac{c}{2m\omega_1} = \frac{c}{c_c} \tag{1.25}$$

The quantity $2m\omega_1$ is known as the critical damping coefficient c_c. In structural engineering it is extremely rare for $\zeta > 1$ so this case will not be considered further. For $\zeta < 1$ the full solution to the differential equation is

$$x(t) = e^{-\zeta\omega_1 t} . X\sin(\omega_d t + \phi) \tag{1.26}$$

The phase angle ϕ depends upon the set time origin and the initial conditions for the motion. X is the amplitude constant for the motion (though this value of displacement may not actually be reached). The damped natural frequency is

$$\omega_d = \omega_1 \sqrt{(1-\zeta^2)} \tag{1.27}$$

In very many real cases $\zeta < 0.1$ and for such small damping ratios

$$\omega_d \approx \omega_1$$

For a system subjected to an impulse the initial conditions at time $t = 0$ are

$$x(0) = 0$$
$$\dot{x}(0) = V = \text{Impulse/mass}$$

and the solution is then

$$x(t) = e^{-\zeta\omega_1 t}X\sin\omega_d t \tag{1.28}$$

where

$$X = V/\omega_d$$

This is a useful result for any problem in which the system is acted upon by a force of short duration compared with the natural period $T = 2\pi/\omega_d$. 'Short' here means one-tenth or less.

If the exponentially decaying vibration can be measured and plotted out (see figure 1.8, for example) then the damping ratio is easily determined from the so-called logarithmic decrement (symbol δ, logdec for short). This is defined as the natural logarithm of the ratio (> 1) of two successive peaks (taken over a whole cycle) and is derived from the decay plot. It can also be usefully derived from the formula

$$\delta = \frac{1}{m}\ln\left(\frac{X_n}{X_{n+m}}\right) \tag{1.29}$$

where X_n is the amplitude of the nth cycle and X_{n+m} is the amplitude of the $(n + m)$th cycle. In figure 1.8, m has a value of 3. It can be shown that

$$\delta \approx 2\pi\zeta \text{ (accurate for } \zeta < 0.1) \tag{1.30}$$

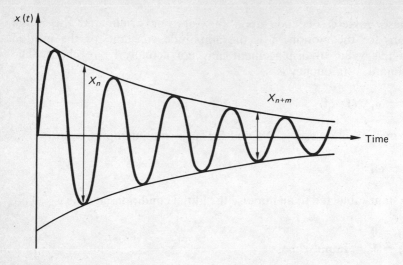

Figure 1.8   Exponential decay of vibration following an impulse

1.2.2   Forced vibration

When a time-varying force $F(t)$ is applied to the system the differential equation of motion of the system is now inhomogeneous:

$$m\ddot{x} + c\dot{x} + kx = F(t) \tag{1.31}$$

This problem has already been solved for the case when $F(t)$ is impulsive. There are three other classes of forcing which may arise:

- Harmonic loading
- Transient loading
- Random continuous loading

A brief account of the first of these is given in the following section. The other two classes of loading may also frequently occur. For example, heel impact on a floor in walking provides a transient loading: acoustic loading on a structure provides random continuous loading. A fuller account of vibration analysis in these latter two cases is given in reference [1].

Harmonic loading

Harmonic loading may be caused in a number of ways, for example, by rotating machinery (as a result of out-of-balance) or by a continuously walking pedestrian. It is characterised by the equation

$$F(t) = F_0 \cos \omega_0 t \tag{1.32}$$

in which

F_0 = Loading or force amplitude
ω_0 = Circular frequency of loading

Ignoring the initial transient motion the continuous steady-state motion which results is

$$x(t) = \frac{F_0}{k} \text{ (DMF) } \cos (\omega_0 t - \phi) \tag{1.33}$$

The phase angle ϕ by which the motion sinusoid lags behind the force sinusoid is termed the *phase lag*. The non-dimensional constant (DMF) is the *Dynamic Magnification Factor* which describes how much greater the displacement is dynamically than it would be under a static load of F_0. Its value is

$$\textbf{(DMF)} = \left\{ \left(1 - \left(\frac{\omega_0}{\omega_1}\right)^2 \right)^2 + 4\zeta^2 \left(\frac{\omega_0}{\omega_1}\right)^2 \right\}^{-\frac{1}{2}} \tag{1.34}$$

Figure 1.9 shows equation 1.34 plotted as dB of magnification against the frequency ratio.

The phase lag ϕ is given by:

$$\tan \phi = \frac{2\zeta \dfrac{\omega_0}{\omega_1}}{\left[1 - \left(\dfrac{\omega_0}{\omega_1}\right)^2 \right]} \tag{1.35}$$

The maximum value of (DMF) occurs when

$$\omega_0 = \omega_1 (1 - 2\zeta^2)^{\frac{1}{2}} \tag{1.36}$$

and this is a condition of *resonance*. If damping is small ($\zeta < 0.1$) then

$$\omega_0 \approx \omega_1$$

and the peak value for (DMF) is

$$\textbf{(DMF)}_{\text{max}} \approx 1/(2\zeta) \tag{1.37}$$

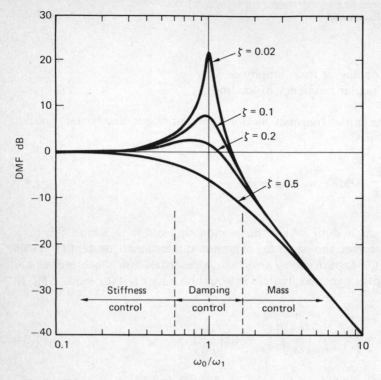

Figure 1.9

Under these conditions the phase lag is

$$\phi \approx 90°$$

1.2.3   Control of forced vibration

The problem often arises of how to control forced vibration. The DMF obviously depends on the physical characteristics of the system; mass, stiffness and damping. It also depends on the ratio of the driving frequency ω_0 to the natural frequency ω_1. The vibration can most obviously be controlled by altering the mass, stiffness or damping. Considering equations 1.33 and 1.34 together it is easy to show for the various ranges of frequency ratio:

(i) for $\omega_0/\omega_1 \ll 1$:

$$|x| = \frac{F_0}{k}$$

Thus, in this region, stiffness controls the motion and would be the physical characteristic that requires to be increased if the motion is to be reduced.

(ii) for $\omega_0/\omega_1 \approx 1$:

$$x = \frac{F_0}{\omega_1 c}$$

It can be seen that damping has a strong influence in this region and is usually the simplest controlling factor. For this reason it is usually termed the damping control region. However, it is worth noting that changes in mass or stiffness will alter ω_1 and may take the problem out of this region. This can have a beneficial effect for harmonic excitation but is of no value if there is excitation across a wide spectrum of frequencies.

(iii) for $\omega_0/\omega_1 \gg 1$:

$$|x| = \frac{F_0}{m\omega_0{}^2}$$

In this region mass is the controlling characteristic.

The control regions are shown on figure 1.9. It is surprising how often vibration control engineers ignore these simple rules. There seems to be, in some, an instinctive feel that stiffness is the factor which must always be increased to solve a problem. This is probably based on an unconscious desire to increase strength. However, it can be seen that this strategy will only work in the lowest of the three ranges indicated on figure 1.9.

1.2.4 Vibration isolation

In many machines the force F_0 in equation 1.32 is generated internally by, for example, out-of-balance of rotating parts, reciprocating parts or internal impacts. The extent to which these forces are transmitted to the supporting structure (the ground in the model of figure 1.7) governs not only the noise which the machine may make but also damage to these supports. It is important therefore to minimise it. The force transmitted to the foundation is, as a result of the attachment of the spring and damper to ground:

$$F_t = c\dot{x} + kx \tag{1.38}$$

The transmissibility T is defined by the equation:

$$T = \left| \frac{F_t}{F_0} \right| \tag{1.39}$$

and this may be evaluated from the harmonic solution to equation 1.31 to be

$$T = \left\{ \frac{1 + 4\zeta^2 \left(\dfrac{\omega_0}{\omega_1} \right)^2}{\left(1 - \left(\dfrac{\omega_0}{\omega_1} \right)^2 \right)^2 + 4\zeta^2 \left(\dfrac{\omega_0}{\omega_1} \right)^2} \right\}^{\frac{1}{2}} \tag{1.40}$$

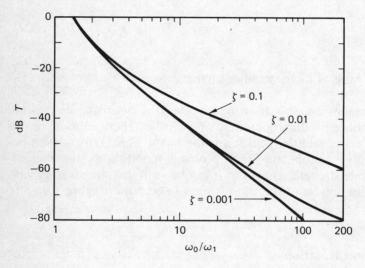

Figure 1.10

This is shown plotted in figure 1.10 in dBs of isolation. The principal result from this analysis is that the natural frequency ω_1 should be very low compared with the exciting frequency ω_0. This will generally lead to the use of the softest isolating spring possible, consistent with being able to carry the service loads without material failure in the spring. An important secondary conclusion is that damping in the isolator degrades its performance. However, damping is often necessary in a machine isolator to limit transient motions on start-up as the machine runs through resonance.

The principle of the isolator is not only used to minimise transmitted forces in machines. It can also be used by the noise control engineer to minimise any transmission of vibrations which may subsequently manifest themselves as noises such as, for example, the transmission of vibration through the structural elements of a building. Another example is the soft mounting material needed for the suspension of microphones in a strong vibration environment to ensure that the microphone only picks up the sound and not the vibration.

1.2.5   Continuous systems and their single degree of freedom equivalents

In the final section we shall briefly consider the vibrations of beams and plates and how their fundamental vibration may be characterised as those of an equivalent SDOF system. Some other continuous systems can also be usefully characterised by an equivalent SDOF system but beams and plates are those most commonly encountered in acoustical engineering. The basis of the analysis of real continuous systems is either (a) the continuum differential equation of motion for the system, or (b) a discrete finite element approximation of it, which can be more or less complex. In both cases the analysis can be reduced, by suitable coordinate transformations, to a problem involving a set of simple oscillators each of which describes one of the characteristic vibrations of the system. This is the basis of the so-called normal mode method and the details of it can be found in good standard textbooks such as references [1] and [2]. In certain circumstances, which occur quite often in practice, only the fundamental mode of vibration is important and so the continuous system can be approximated by an equivalent SDOF system. The circumstances in which this approximation works well are:

(a) when the spatial distribution of forces is reasonably uniform;
(b) when the rate of force application is less than or equal to the fundamental natural frequency;
(c) when the two lowest natural frequencies are not close in value.

In other circumstances, consideration must be given to a more rigorous analysis involving the use of more modes of vibration, but this is beyond the scope of this account.

The characteristic equivalent values, which are summarised in this section for mass and stiffness, together define the fundamental natural frequency for the system. They are derived by the method of generalised coordinates (see references [1] and [2]) or by some close approximation to it.

For the substitute SDOF system to be equivalent, its properties have to be chosen such as to yield *the same fundamental natural frequency* and *the same displacement amplitude* – under the substitute force $F(t)$ – as the original system. The first step is to choose an appropriate reference point, the motion of which is $x(t)$ in equation 1.22. For beams this is usually chosen to be either at the point of maximum displacement or at the centre. For the plates indicated, the reference point is at the centre. In what follows the approximate equivalent parameters for the substitute SDOF systems are called the 'generalised' parameters:

Φ_L is the factor by which to multiply either the uniformly distributed load or the point load at the centre in order to determine the generalised force magnitude. The time variation of the force is unaltered. This gives $\tilde{F}(t) = \Phi_L F(t)$.

Φ_M is the factor by which to multiply either a total uniformly distributed mass or a point mass at the centre in order to determine the generalised mass. This may be expressed $\tilde{m} = \Phi_M m$.

The effective stiffness k is the static stiffness at the reference point under the action of the relevant type of load. The generalised stiffness is given approximately by $\tilde{k} = \Phi_L k$.

If damping needs to be taken into account, as is sometimes the case in forced vibration analysis, the correct value for the damping ratio ζ must be used. This may either be estimated on the basis of past experience or, in some cases, it may be directly measured.

The approximate equivalent SDOF equation of forced motion is then:

$$\tilde{m}\ddot{x} + 2\zeta\sqrt{(\tilde{m}\tilde{k})}\,\dot{x} + \tilde{k}x = \tilde{F}(t) \tag{1.41}$$

and this can be solved in the manner of the preceding section. The natural frequency is given by

$$\omega_1 = \sqrt{\left(\frac{\tilde{k}}{\tilde{m}}\right)} = \sqrt{\left(\frac{\Phi_L k}{\Phi_M m}\right)} \tag{1.42}$$

Tables 1.3 and 1.4 (courtesy of Prof. H. Bachmann, ETH, Zürich) gives approximate values for the generalised parameters for beams and plates respectively.

In these tables (see pages 30 and 31) the product EI is the flexural stiffness of the element. E is the modulus of elasticity for the material. For some materials, such as plastics, this modulus has a different value under dynamic conditions from its static value. In practice the E-value for plates is modified compared with beams of the same material, because of the plane stress conditions. The effective modulus is then given by the expression

$$E/(I - \nu^2)$$

where ν is the Poisson's ratio for the material, and I is the second moment of area for the cross-sectional shape of the beams. For plates, I_0 is the second moment of area for a unit width of plate and is given by

$$I_0 = t^3/12$$

t being the thickness of the plate.

REFERENCES

[1] W. T. Thompson, *Theory of Vibration*, 3rd edn, Prentice-Hall (1988).

[2] R. W. Clough and J. Penzien, *Dynamics of Structures*, McGraw-Hill (1975).

Table 1.3 Equivalent substitute SDOF parameters for single span beams with various support and load conditions (see explanation in the text).

Loading and support conditions	Load factor Φ_L	Mass factor Φ_M		Effective beam stiffness k
		Central lumped mass	Uniformly distributed mass	
	0.64	—	0.5	$\dfrac{384\ EI}{5\ l^3}$
	1.0	1.0	0.494	$\dfrac{48\ EI}{l^3}$
	0.578	—	0.448	$\dfrac{185\ EI}{l^3}$
	1.0	1.0	0.43	$\dfrac{107\ EI}{l^3}$
	0.533	—	0.406	$\dfrac{384\ EI}{l^3}$
	1.0	1.0	0.383	$\dfrac{192\ EI}{l^3}$

Table 1.4 Approximate equivalent SDOF parameters for rectangular plates with various support conditions (uniformly distributed mass and load only); the factors have been derived from simultaneous excitation of more than one mode of vibration as a result of impact loading and are therefore only approximate SDOF equivalents. For low values of the aspect ratio *a/b*, treat as a beam across the short dimension (*a*) using table 1.3

Support conditions	$\dfrac{a}{b}$	Load factor Φ_L	Mass factor Φ_M	Effective plate stiffness k
$\dfrac{b}{2} \leqslant a \leqslant b$ simply supported all round	1.0	0.45	0.31	$\dfrac{271\,EI_0}{a^2}$
	0.9	0.47	0.33	$\dfrac{248\,EI_0}{a^2}$
	0.8	0.49	0.35	$\dfrac{228\,EI_0}{a^2}$
	0.7	0.51	0.37	$\dfrac{216\,EI_0}{a^2}$
	0.6	0.53	0.39	$\dfrac{212\,EI_0}{a^2}$
	0.5	0.55	0.41	$\dfrac{216\,EI_0}{a^2}$
$\dfrac{b}{2} \leqslant a \leqslant b$ clamped all round	1.0	0.33	0.21	$\dfrac{870\,EI_0}{a^2}$
	0.9	0.34	0.23	$\dfrac{798\,EI_0}{a^2}$
	0.8	0.36	0.25	$\dfrac{757\,EI_0}{a^2}$
	0.7	0.38	0.27	$\dfrac{744\,EI_0}{a^2}$
	0.6	0.41	0.29	$\dfrac{778\,EI_0}{a^2}$
	0.5	0.43	0.31	$\dfrac{886\,EI_0}{a^2}$

2. The human effects of noise: criteria and units

What is it that roareth thus?
Can it be a Motor Bus?
Yes, the smell and hideous hum,
Indicat Motorem Bum.
A. D. Godfrey

2.1 INTRODUCTION

The ear is rather like a microphone. A diaphragm (the eardrum) moves in response to pressure waves, and the motion is sensed and converted to an electrical signal by nerve endings. The signal is passed to the brain, where it gives rise to the sensation of sound. The ear, like a microphone, creates an electrical analogue of the sound. The signal processing carried on in the brain enables us to discriminate between different tones, to recognise speech, to enjoy music, and even (sometimes) to identify a speaker through the distortion of a telephone.

Unlike a good-quality microphone however, the human ear does not possess uniform sensitivity across the audio frequency range. It is much more sensitive to some frequencies than others (notably those involved in speech, for good evolutionary reasons). The frequency response of the ear is not constant, but changes slowly as we grow older with a decrease in the ability to hear high frequencies. Rapid reductions in sensitivity may occur if we are exposed to damaging noise levels. Studies of the ear's sensitivity have led to the growth of audiometry, the science of measuring hearing. Audiometric measurements are routinely carried out to ascertain the extent of hearing loss due to ageing, disease and damage.

The ear is also susceptible to a phenomenon known as masking. Speech which is perfectly audible in quiet surroundings may become unintelligible as the ambient noise level increases. Masking is a complicated effect which depends not only on the relative pressure levels of two sounds, but also on their frequency content. A pure tone is more effectively masked by a second pure tone of similar frequency than by a pure tone of quite different frequency. In general, low-frequency noises mask high-frequency noises more than high-frequency noises mask low frequencies. This is why it is difficult to get top quality sound from a car stereo, no matter how much money is invested, as car body resonances usually occur at the low end of the audio spectrum and cause low-frequency background noise.

The microphones, amplifiers and measuring equipment used to estimate noise levels are designed to have a flat frequency response. However, the purpose of noise measurements is often to assess the effect of a sound on people. It is therefore necessary to make the measuring equipment act as much like an ear as possible. Weighting circuits are used for this purpose which attempt to emulate the ear's sensitivity at different frequencies. Three internationally agreed weighting functions (known as A, B, and C) were introduced, to be applied at different sound pressure levels (*SPLs*) as discussed later in this chapter. However, it has been found that for many common noises the A-weighting is sufficient at all sound pressure levels (the units used are dBA). Other weighting functions are used in special circumstances, notably the D-weighting for measuring the noise produced by a single aircraft flying past, and the SIL weighting, used to assess speech interference.

The annoyance caused by a sound is partly, but not entirely, due to its perceived loudness. Loudness is well understood, and although the measurement techniques are still subject to refinement they are well established in the appropriate standards. The annoyance caused by a sound is not so well defined. In trying to predict the likely reaction to a noise there is obviously a need for some scale which relates the subjective response of a human community to a readily measurable feature of the noise. A threshold can then be defined above which exposure to the noise becomes unacceptable. A number of such rating scales have been defined, some requiring knowledge of the spectral content of the noise, others requiring knowledge of the noise history, and others again which are based on combinations of the two.

The intention in this chapter is to introduce the topics mentioned above and to give some suggestions for further reading. Considerations of space preclude more than a brief discussion of the human reaction to sound, which is often termed psychoacoustics. The subject is of some complexity but great interest, and is well worth further study.

2.2   THE EAR

The human ear is shown in section in figure 2.1. Sound pressure waves are collected by the outer ear and cause deflections of the eardrum. The atmosphere exerts much more pressure on the eardrum than any sound, and if there were no compensatory mechanism it would rupture. However, a small passage (the eustachian tube) connects the inner ear with the mouth, and equalises the atmospheric pressure on both sides of the

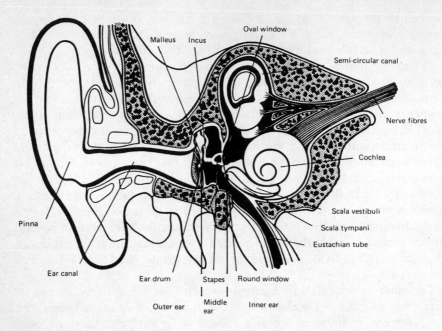

Figure 2.1   The human ear in section (courtesy of Bruel & Kjaer Ltd)

eardrum. The eustachian tube operates slowly and is only opened during the action of swallowing, so the ear is unaffected by slow meteorological changes in atmospheric pressure. If the atmospheric pressure changes quickly, as happens in a rapid descent in an aircraft, the equalisation mechanism may not operate fast enough and pain is felt. If the eustachian tube opened directly to atmosphere the pressure compensation would be instantaneous but the ear would not work, since not only atmospheric changes but also sound pressure would be compensated for.

A system of small bones in the middle ear (the malleus, incus and stapes, collectively known as the ossicles) act as mechanical levers to amplify the deflections and pass them to the oval window. This separates the air-filled middle ear from the liquid-filled inner ear. Deflections of the oval window cause pressure changes in the inner ear, which acts as a transducer, conveying acoustic information to the brain.

The inner ear contains two systems, the cochlea, which is the transducer concerned with hearing, and the semi-circular canals, which operate as tilt sensors and allow us to balance.

The liquid-filled cochlea consists of a long tube-like cavity rolled into a spiral. The cavity is divided into two longitudinal canals by the basilar membrane, which extends the length of the cochlea except for a small gap (the helicotrema) at the far end. Figure 2.2 shows the unrolled cochlea.

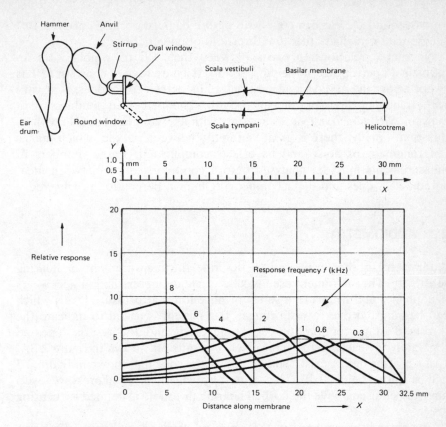

Figure 2.2 *Longitudinal section of the cochlea showing the positions of response maxima (courtesy of Bruel & Kjaer Ltd)*

When the oval window moves in response to an acoustic event it creates a pressure pulse, which travels along the upper canal, passes through the helicotrema, and returns by the lower canal to the round window, which is deflected. During the passage of a pressure wave the disturbance distorts the basilar membrane, on whose surface there are sensitive hairs. Deflection of these hairs produces the electrical nerve impulses sensed by the brain.

Different parts of the basilar membrane are sensitive to different frequencies. The maximum response to high frequencies occurs near the oval window, and that at low frequencies takes place near the helicotrema. Figure 2.2 summarises the sensitivity of different parts of the basilar membrane to a number of pure tones.

The structure of diaphragms, levers and sensitive hairs described above allows the ear to cope with a frequency range from about 20 Hz to 16 kHz, and with a dynamic *SPL* range of about 120 dB. In terms of intensity, the loudest sound the ear can cope with before damage occurs is around 10^{14} times greater intensity than the threshold of perception.

No simple relationship exists between the *SPL* of a noise and an individual's perception of the same sound. If a pure tone of constant *SPL* is swept across the audio frequency range, its perceived loudness will vary with frequency in a non-linear fashion. The perceived loudness of a constant *SPL tone burst* varies not only with its frequency but also with its duration. Finally, there is great variability between people, which makes the treatment of noise and its effects complicated. Large numbers of measurements must be analysed before good correlation between measured *SPL* values and the accompanying human perception is achieved.

2.3 AUDIOMETRY

Audiometry is the term used to describe the measurement of hearing sensitivity. The instrument used is called an audiometer. The subject wears headphones and listens to a series of pure tones at pressure levels which can be adjusted over a wide range. The patient is asked to indicate the threshold of perception for each tone used, and the test is repeated separately for each ear. A typical set of results are shown in figure 2.3.

If a hearing defect exists, some or all of the solid line shown in figure 2.3 will be shifted below 0 dB. The amount by which the sound pressure levels have to be raised above the normal hearing threshold is defined as 'hearing loss'.

The ambient noise level in the room in which an audiometer is being used is obviously important. Although the subject wears earphones for the test, enough noise may penetrate to mask the test tones and give an artificially high figure for hearing loss. For this reason, hospitals generally use a soundproof room for audiology, in which the background noise level is low enough not to interfere with the measurements.

2.4 LOUDNESS

The perceived loudness of a sound depends largely on its SPL and frequency content. Several curves of equal loudness have been proposed. Each set of curves is the result of a large number of psycho-acoustical experiments, and is therefore valid only for the particular test conditions used. For example, the sound source may be a pure tone or a frequency band, and the subject may be in a free or reverberant field. The curves are obtained by averaging and smoothing data obtained from large numbers of people. Normally the test subjects are of similar ages to reduce variability. Figure 2.4 shows a frequently used set of loudness data known as the

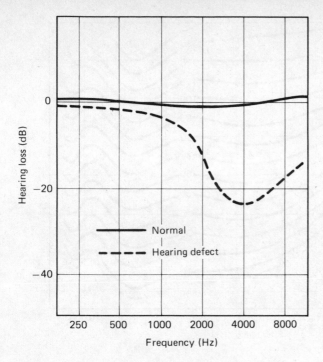

Figure 2.3   Typical audiometer result

Fletcher–Munson curves. They are obtained under the following conditions:

(1) the sound source (a pure tone) is directly ahead of the listener;
(2) the sound reaches the listener in the form of a free progressive plane wave (that is, the listener is in the free field);
(3) the sound pressure level or *SPL* is measured in the *absence* of the listener;
(4) both ears are used;
(5) the test subjects are in the age group 18 to 25 years inclusive, and have normal hearing.

The Fletcher–Munson curves are obtained as follows. The tone to be judged is played to the subject, who has the task of adjusting the amplitude of an alternately heard 1 kHz reference tone until it has the same perceived loudness.

The reference tone has a pressure level of 2×10^{-5} N/m^2, and defines the zero point on a loudness scale. The unit of perceived loudness is the phon. The reference *SPL* (2×10^{-5} N/m^2) also defines the 0 dB point on

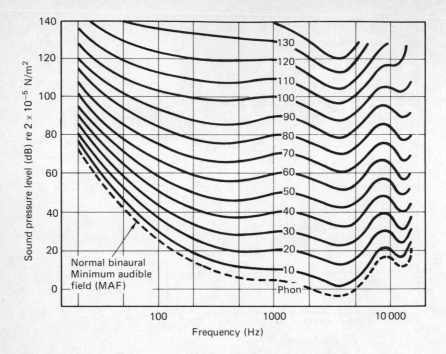

Figure 2.4   Fletcher–Munson equal loudness contours for pure tones

the vertical axis. As figure 2.4 shows, at frequencies other than 1 kHz the ear's sensitivity varies. Only at the reference frequency is the loudness of a sound in phons equal to its *SPL* in dB.

Figure 2.4 allows us to relate the measured *SPL* of a sound to its apparent loudness as perceived by an average listener. For example, suppose the *SPL* of a tone at 3 kHz is 72 dB. Its loudness will be 80 phons. Similarly, an *SPL* of 72 dB at 9 kHz will have a loudness of around 65 phons.

It can be seen from figure 2.4 that the variation in the ear's sensitivity is greatest at low frequencies. At 40 Hz an *SPL* of 52 dB is required to create a loudness of 10 phons, compared with an *SPL* of 10 dB at 1 kHz.

When measuring the likely annoyance caused by a noise, it is convenient to be able to convert the noise meter reading from *SPL* in dB to some quantity related to the ear's sensitivity. Most sound-measuring equipment is provided with one or more weighting circuits for this purpose. Weightings are discussed in detail in the next section of this chapter.

The loudness of more complex sounds such as band-limited noise can also be determined by the comparison method. The procedure is more complex than that used for pure tones, as a correction for the effects of masking must be made. The method generally used enables the loudness of complex broadband noises (which may include pure tones) to be calculated for diffuse and free-field conditions. This method forms the basis of an internationally accepted standard loudness calculation procedure (see reference [1]).

2.5 WEIGHTING FUNCTIONS

Most sound-measuring instruments contain built-in weighting networks. These are electronic filters with a frequency response which is approximately equivalent to the equal loudness curves of figure 2.4. Since the curves are not parallel, different weightings are applied at different *SPL*s. The A weighting is derived from the 40 phon contour, and is intended for use for *SPL*s below 55 dB. The B- and C-weightings follow the 70 and 100 phon contours respectively. The B-weighting was intended for use between pressure levels of 55 and 85 dB, and the C-weighting for higher levels. All three weighting functions are shown in figure 2.5.

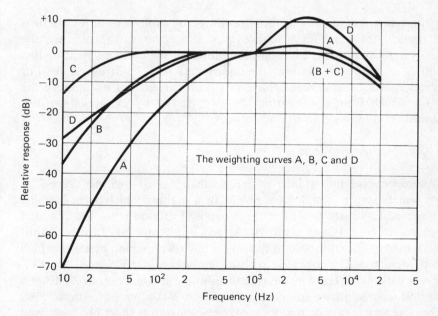

Figure 2.5   *A-, B-, C- and D-weighting functions*

Unfortunately, it was found that in practice this arrangement did not give reliable loudness estimates. However, for many common noises the simple use of the A-weighting at all pressure levels gives readings which are reasonably well correlated with perceived loudness. Since measurements in phons can only be achieved under carefully controlled laboratory conditions as described earlier, their use for practical measurements is rare. Measurements of loudness and exposure criteria are almost always given in dBA, in other words as *SPL* measurements weighted by the A-function.

Other weighting functions are sometimes also applied for specific measurements. A D-weighting is used for measuring the loudness of aircraft during take-off and in flight. A combination of B and C is sometimes used for high sound pressure levels. Both of these are shown on figure 2.5.

2.6 NOISE-RELATED HEARING LOSS

If a noise is excessively loud it can damage the structure of the ear, resulting in a temporary or permanent loss of hearing. Noise-induced hearing loss occurs in two ways:

(1) Trauma

High-intensity sounds such as those produced by explosions or jet exhausts can rupture the eardrum. Damage may also occur to the ossicles, the sensory hairs covering the basilar membrane, or the cochlea. Hearing loss resulting from trauma is always sudden and may be irreversible. For obvious reasons, experiments have not been carried out to measure the *SPL* at which traumatic hearing loss occurs, but by extrapolation it is assumed to take place at around 150 dB.

(2) Chronic hearing loss

Of more general importance is the gradual loss of hearing caused by persistent exposure to high noise levels, such as those which occur in many factories and workshops. If the ear is exposed to loud noise for a limited time, a short-term loss of sensitivity known as a temporary threshold shift (TTS) occurs. The threshold of hearing is the lowest sound pressure which can be detected. For a person with normal hearing it is about 2×10^{-5} N/m^2 or 0 dB at 1 kHz. The hearing threshold can rise by up to 20 dB at 4 kHz (the most sensitive frequency, see figure 2.4) after exposure to loud noise. The loss is usually temporary if the exposure is short however, and the ear recovers its sensitivity fairly rapidly.

After cessation of a noise loud enough to cause TTS (but not permanent damage), the ear recovers its sensitivity in a characteristic pattern. The TTS first decreases, and then increases to a maximum about 2 minutes after the noise has stopped. This is called the bounce effect. After 2 minutes the TTS decreases until the original sensitivity is recovered.

Unfortunately many people work in noisy environments for 8 or more hours a day, all year round. Under these conditions the hearing loss is no longer temporary. A permanent hearing loss develops over a period of years which may be severe enough to make normal conversation difficult. The damage is irreversible, and no amount of rest gives any significant recovery (see table 2.4, page 49).

The form of noise-induced hearing loss revealed by audiometry is almost independent of the characteristics of the sound which caused the damage. The greatest reduction in sensitivity occurs at around 4 kHz. As exposure times increase the sensitivity loss at 4 kHz becomes greater, and the damage extends to lower frequencies.

2.7   NOISE EXPOSURE LIMITS AND NOISE UNITS

Individuals vary in their reaction to noise. For some types of noise source, special measurement units are used as a result of either the special characteristics of that type of noise, or of the special environment in which it can be a nuisance. It is necessary therefore to have a range of admissible conditions rather than an absolute standard. For example, most people find the noise level in a railway station acceptable, while they would find exactly the same sounds intolerable in a living room. The permissible level of a noise is a function of its loudness, frequency content, the exposure time, and any interference which it causes with other activities such as speech or listening to music. Most noise criteria are related to loudness and frequency content, not least because these are straightforward to measure. However it is much more difficult to quantify the loss of information contained in a sound and few criteria attempt to measure this.

For steady background noise the most common approach is to use the dBA scale, or the NC or NR noise rating curves. For variable noise a statistical approach is adopted, in which the noise levels exceeded for a certain percentage of the time are estimated. An example is the L_{10} scale, in which the noise level in dBA exceeded for 10 per cent of the time is measured. An alternative approach to variable noise is to assess the amount of noise energy received over a given period. This gives a 'noise dosage' figure which can be related to an equivalent continuous sound level (L_{eq} in dBA).

2.8   THE NOISE CRITERION (NC) AND NOISE RATING (NR) CURVES

NC curves specify the maximum sound pressure levels permissible in each octave band. They take the form of a family of curves derived from the equal loudness contours shown in figure 2.4. The rating of a noise is determined by the point on its spectrum which comes highest relative to the NC curves. Figure 2.6 shows the NC curves.

The NC curves have been found to be unsatisfactory in some respects and are only usable over a limited spectral and loudness range. Noise rating (NR) curves were subsequently developed to overcome these objections and are shown in figure 2.7.

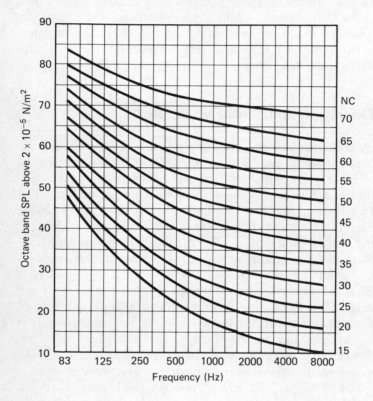

Figure 2.6   NC curves

2.9 L_{10} and L_{90}

Many noises vary in loudness. Examples are traffic and aircraft noise, and occupational noise in some factories. The reaction of people to variable noise is strongly related to its range. Figure 2.8 shows a typical record taken from measurements of traffic noise in a building adjacent to a road. The L_{90} line is the sound level (in dBA) which is exceeded for 90 per cent of the time. The L_{10} line is the sound level which is exceeded 10 per cent of the time. The region between L_{10} and L_{90} is referred to as the noise climate.

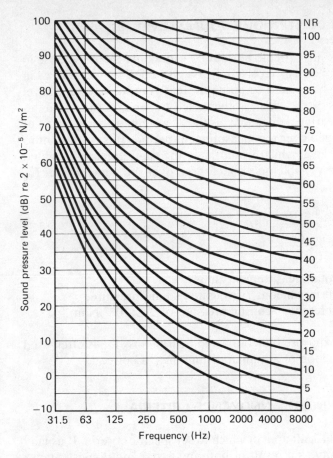

Figure 2.7   NR curves

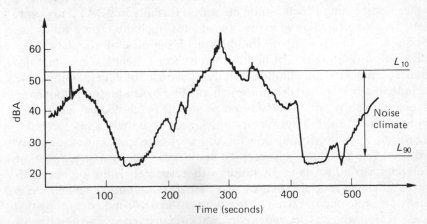

Figure 2.8 L_{10}, L_{90} and noise climate estimates on a typical noise record taken from roadside traffic measurements

2.10   EQUIVALENT CONTINUOUS SOUND LEVEL (L_{eq} dBA)

Another way of assessing variable noise is by estimating the noise dose in terms of the amount of energy received over a given period. The energy in a variable noise may be related to an equivalent continuous sound level over the period of interest, which is then used to set exposure limits to prevent permanent hearing damage. L_{eq} is the A-weighted energy mean of the noise level, averaged over the measurement period. It is equivalent to the continuous noise level which would have the same total A-weighted acoustic energy as the variable noise, measured over the same period. L_{eq} is defined as

$$L_{eq} = 10\log_{10}\frac{1}{T}\int_0^T \left\{\frac{P_A(t)}{P_0}\right\}^2 dt \qquad (2.1)$$

where T is the total measurement time,

$P_A(t)$ is the instantaneous A-weighted sound pressure,

P_0 is the reference acoustic pressure of 2×10^{-5} N/m^2.

As well as setting exposure limits to avoid hearing loss L_{eq} is often used for estimating the reaction of a human community to noise.

2.11   COMMUNITY NOISE ANNOYANCE CRITERIA

This is a much more difficult area in which to set specific criteria. It usually arises in the community as a result of both industrial and domestic noise sources (and occasionally others). Nuisance will be a function of loudness, and this is something which can be measured (usually in dBA). However, nuisance is also a function of many other factors including some which are a response of the psyche of the hearer. Examples of the latter are predisposition (getting out of bed the wrong side), health, economic tie, prejudice, animosity and emotional state. These are, of course, impossible to include in any standard criteria which must be based on the response of an average, healthy and well-balanced person. If psychological factors are excluded there are still other factors which need to be taken into account. The easiest way of illustrating how this can work is to give a brief summary based loosely on a range of accepted standards. The broad purpose of these standards is to assess a noise with respect to interference with working efficiency, social activities or personal tranquillity. The method is basically a simple one of taking a measurement in dBA and comparing this with a criterion which is dependent on local circumstances. The assessment then depends on the degree to which the measured noise exceeds the set criterion. If it is necessary to take noise-reducing steps then further measurements may be made in octave bands and these measurements can be compared with NR curves, see figure 2.7. These curves are an equivalent-frequency-banded version of the single figure criterion. They enable the frequency bands containing the problem to be identified and appropriate noise control steps may then be taken.

The noise is first measured in dBA (or L_{eq} if the noise is intermittent or variable) and then corrected for the character of the noise. Thus the measured noise value is converted to an equivalent steady noise level. The corrections might be as follows:

(a) if the noise contains impulsive peaks, measure the average peak level and add 5 dB;
(b) if a pure tone is perceptible, add 5 dB.

This procedure takes account of factors which are a function of the character of the noise itself and gives L_r, the rating sound level.

The basic criteria of acceptable sound levels in dBA are as shown in table 2.1. These criteria were originally derived from research in The Netherlands. Personal reaction does vary considerably, so some caution is needed in making universal applications of these criteria.

Table 2.1 Acceptable sound levels (dBA).

1. Broadcasting studio	15
2. Concert hall, theatre	20
3. Class room, music room, TV studio, conference room	25
4. Bedroom	35
5. Cinema, hospital, church, courtroom, library, living room	40
6. Private office	50
7. Restaurant	55
8. Sports hall, office with typewriters	55
9. Workshop	65

Corrections should be made to these values for the time of day, but these apply only to items 4, 5, 6, 7 and might be:

(a) Day time only   0 dBA
(b) Evening   −5 dBA
(c) Night time   −10 dBA

In addition, a correction for type of district can be made when the criterion is to be determined for residential premises (that is, domestic noise problems associated again with items 4, 5, 6 and 7 in the basic list). These corrections are shown in table 2.2.

Table 2.2

Type of district	Corrections to basic criterion (dBA)
Rural residential, hospital area, recreation	0
Suburban residential, little road traffic	+ 5
Urban residential	+10
Urban residential with workshops, business or main roads	+15
City (business, trade)	+20
Predominantly industrial area (heavy industry)	+25

Assessment of the problem can now be made by comparing the corrected sound level with the corrected noise criterion according to table 2.3

Table 2.3

Amount in dBA by which the rating sound level L_r exceeds noise criterion	Estimated community response	
	Category	Description
0	None	No observed reaction
5	Little	Sporadic complaints
10	Medium	Widespread complaints
15	Strong	Threats of community action
20	Very strong	Vigorous community action

If there is a chance of complaint, as predicted by table 2.3, then a frequency analysis may be desirable using the NR curves, figure 2.7. In addition an individual company or noise consultant may wish this procedure to be adopted for the purposes of noise control design. The approach is as follows. First, sound pressure levels are measured in octave bands. Each SPL is plotted on figure 2.7, and compared with the NR curve which has a numerical value given by the calculated criterion. If the measured spectrum exceeds the appropriate NR curve by more than 5 dB in any octave, it is deemed to be 'difficult to accept'; if by more than 10 dB, 'unacceptable'.

2.12   AIRCRAFT NOISE MEASUREMENTS: UNITS AND CRITERIA

The annoyance caused by aircraft is the subject of much research in acoustics. This generally falls into two areas. These are the measurement of single events, such as one aircraft flying past, and assessments of the long-term noise exposure of overflown populations. Studies have shown that annoyance from a single flyover is closely related to peak level, and that a suitable unit for single-event measurement is the Perceived Noise Level (PNL or PNdB). For long-term noise exposure assessments, the number of flyovers (N) must also be taken into account. The unit used until recently for this purpose is the Noise and Number Index (NNI), where

$$\textbf{NNI} = \textbf{(Average Peak PNdB)} + \textbf{(15log}_{10}\textbf{\textit{N})} - \textbf{(80)}$$

A new approach has recently been introduced [2], in which NNI assessment is replaced by measurements of the L_{eq} noise dose (see section 2.10) averaged over the 16-hour period from 0700 to 2300 hours. The approximate ratings for annoyance using both schemes are as follows:

NNI	L_{eq}(16 hour)(dBA)	Response
35	57	Little
45	63	Moderate
55	69	Significant

A typical aircraft noise-measuring arrangement consists of a sound level meter (see chapter 6) fitted with a remote microphone. The microphone is positioned about a metre above an acoustically hard surface and is placed at grazing incidence to the flight path. It is important that no obstructions are present within the microphone's cone of sensitivity. This is why a remote microphone is generally used, since the noise meter itself may constitute an obstruction. Obviously the operator is also an obstruction, and he or she should move well away from the microphone during measurements.

Approximate instantaneous values of peak PNL can be obtained from A- or D-weighted sound meter readings by applying the following corrections:

$$\textbf{PNL (dB)} = \textbf{dBA} + \textbf{13}$$
$$\textbf{PNL (dB)} = \textbf{dBD} + \textbf{7}$$

An unweighted recording from a noise meter can yield much more information about the flyover if a full Effective Perceived Noise analysis is undertaken. This may be done in several ways, with the method chosen depending on the quantity of data and the application of the final results.

The simplest approach is to weight the signal (using A or D) and plot it on a graphic level recorder. This provides a time history of the flyover from which approximations to the effective perceived noise level L_{EPN} may be calculated. However, information about the frequency content of the noise is neglected in this approach.

Further analysis is necessary if more accurate values of the effective perceived noise level L_{EPN} are required. A description of the process is beyond the scope of this book, but in essence it consists of summing the noisiness values obtained in $\frac{1}{3}$ octave bands, and using the result to give an L_{EPN} value according to a prescribed formula. This value is then compared with corrected acceptable noise levels as described in the section on community noise annoyance criteria. A full description of the process and comparisons of various aircraft noise rating schemes may be found in references [3] and [4].

2.13   HEARING DAMAGE RISK CRITERIA

Occupational hearing impairment was first noted amongst blacksmiths in 1830 [4]. For steady noise a good correlation exists between hearing damage risk and simple A-weighted sound level measurements, and the dBA unit is now universally used for rating continuous noise. Simple noise meters can be used to make measurements adjacent to any machine or workplace, and an instantaneous assessment made of whether the operator needs to take precautions such as the use of ear defenders. Most standards (such as ISO 1999) quote a maximum peak noise level which should not be exceeded. However, a more important concept is the maximum permissible noise dose, which takes account of both the level of noise and its duration. The dose is the A-weighted equivalent noise level L_{eq} to which someone may be subjected for up to 40 hours a week, before incurring a significant risk of permanent hearing loss. The allowable dose varies slightly between standards, but is usually of the order of 85–90 dBA. In the UK for example, the latest Health and Safety Executive regulations [4] require employers to provide risk information and to make hearing protection available when the noise level is between 85 dBA and 90 dBA. At 90 dBA and above, hearing protection is mandatory.

The allowable dose is referred to as the criterion or 100 per cent noise dose. The advantage of this approach is that 100 per cent always represents the criterion dose whatever the measurement duration and however the dose is accumulated. It is permissible to spend time in a noisy environment at more than 85 or 90 dBA, as long as this is followed by enough time spent in a quieter area to bring the overall dose within the equivalent of 85 dBA for 40 hours a week or 8 hours a day specified by the standard.

The above discussion is valid for continuous and fluctuating noise, but may not be applicable to impulsive noise consisting of transients with less than 1 second's duration. This topic is the subject of much research, and it appears that the harmfulness of impulsive sound depends on a number of factors such as peak pressure, duration, rise time and pulse shape [4]. Most standards only consider the first of these, or an easily measured approximation to it. The latest UK regulations (reference [5]) give some guidance on the treatment of impulsive noise.

Table 2.4 shows the type of information which enables a criterion to be defined. This information, obtained from UK statistics (1981), shows the number of people likely to suffer significant hearing damage at different levels of exposure. Thus a person subjected to 90 dBA, eight hours per day for a working lifetime has an 11 per cent chance of suffering a 50 dB hearing loss (and this is serious damage, which the man in the street would call deafness). The L_{eq} figure used in this table is the energy averaged dBA for an eight hour working. Thus, for example, 93 dBA for only 4 hours would give the same L_{eq} of 90.

Table 2.4 Percentages of persons likely to suffer a 50 dB hearing loss.

Level of exposure L_{eq}	Lifetime exposure (per cent)	10 years' exposure (per cent)
100	32	17
90	11	5
80	3	1

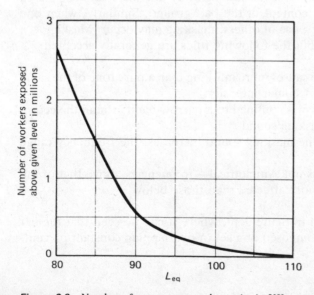

Figure 2.9   Number of persons exposed to noise in UK manufacturing industry (1981)

From this information, one might wish to set the criteria below 80 L_{eq} so as to avoid the problem. However, this may be impracticably costly for industry. Figure 2.9 shows the UK situation as it existed in 1981. As a result, a criterion of 90 dBA for the L_{eq} is widely accepted throughout the world. There is obviously some risk associated with this figure and, in the UK as we have seen, the 1990 HSE regulations require action to be taken when the L_{eq} exceeds 85 dBA.

2.14 MASKING

> *Rats . . .*
> *Made nests inside men's Sunday hats,*
> *And even spoiled the women's chats,*
> *by drowning their speaking,*
> *with shrieking and squeaking,*
> *in fifty different sharps and flats.*
> R. Browning

We are not often exposed to sound containing only one frequency. Most 'real' noise contains energy over a range of frequencies. Even when a single note is played on a musical instrument, the resulting sound will contain frequencies other than the fundamental. These additional frequencies are called harmonics, and it is variation in the harmonic content that distinguishes, say, middle C on a piano from middle C on a violin.

When an information-bearing sound such as speech is being heard in the presence of background noise, the intelligibility or otherwise depends on the level and frequency content of the background. Similarly, when one note is played in the presence of others, masking may occur. Masking is a complex phenomenon, but the following rules are generally accepted:

(1) a narrow-band noise causes more masking than a pure tone of the same centre frequency and similar intensity;
(2) at low levels the effect is confined to a narrow band around the centre frequency of the masking sound;
(3) as the loudness of the masking sound increases, the frequency range affected increases;
(4) the masking effect is not symmetrical – frequencies above that of the masking noise are more affected than those below.

These effects are shown in figure 2.10, where it can be seen that the ear acts as though it were composed of a set of overlapping constant percentage bandwidth filters.

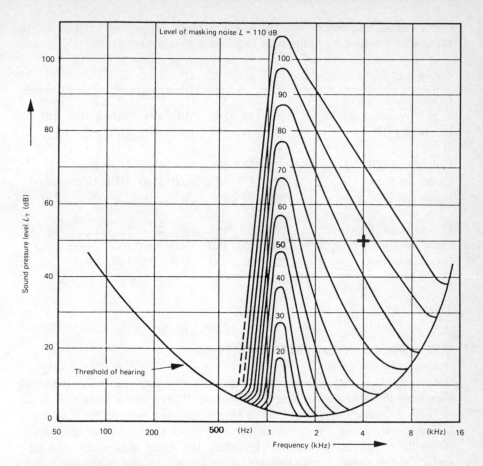

Figure 2.10   Masking effect of narrow band noise at 1.2 kHz. A 50 dB 4 kHz tone (marked +) can be heard if the masking sound level is 90 dB, but is masked at 100 dB (courtesy of Bruel & Kjaer Ltd)

It has been found that as the bandwidth of a masking sound is increased the masking increases, but only up to a critical value, beyond which no further masking occurs. The existence of a critical bandwidth means that only energy close to the frequency of the sound to be masked contributes significantly to the effect. Reference [4] shows that a critical band corresponds to a distance of about 1.3 mm along the basilar membrane (see figure 2.2), and bears a direct relationship to the response maxima along it. The critical distance of 1.3 mm is defined as 1 Bark. The critical bands and their associated bandwidths are given in table 2.5. Over a wide range and especially at high frequencies, the critical bandwidth is around 23 per cent of the centre frequency or $\frac{1}{3}$ of an octave. This fact is often used to justify the use of $\frac{1}{3}$ octave band analysis in noise measurement.

Table 2.5 Critical bands.

Critical band (Bark)	1	2	3	4	5	6	7	8
Centre frequency (Hz)	50	150	250	350	450	570	700	840
Bandwidth (Hz)	100	100	100	100	110	120	140	150
Critical band (Bark)	9	10	11	12	13	14	15	16
Centre frequency (Hz)	1000	1170	1370	1600	1850	2150	2500	2900
Bandwidth (Hz)	160	190	210	240	280	320	380	450
Critical band (Bark)	17	18	19	20	21	22	23	24
Centre frequency (Hz)	3400	4000	4800	5800	7000	8500	10500	13500
Bandwidth (Hz)	550	700	900	1100	1300	1800	2500	3500

2.15   SPEECH INTERFERENCE CRITERIA

Speech intelligibility is important because many activities rely on verbal communication to a greater or lesser extent. There are two main methods of verbal communication, face-to-face conversation and the use of a telephone or other electromechanical system. Communication systems normally compress speech into less than the usual bandwidth and may reduce the dynamic range, so alternative analysis methods are used.

The frequency range and acoustic power of speech vary widely. Consonants in normal speech have a power of around 0.03 μW, while shouting can produce up to 2 mW. The information content of speech is mainly carried by high-frequency low-energy consonants. Noise has the greatest masking effect if it contains significant power above 500 Hz. In addition to masking the perception of speech is affected by factors such as reverberation, signal clipping (that is, reduced dynamic range), and voice quality.

The requirement for clarity of speech communication is put at risk by the presence of background noise and an example of how this can sometimes be even a safety hazard is shown in the swimming pool case study in chapter 7. The main frequency band for speech is 500 Hz to 4 kHz and speech levels can vary from a whisper up to energetic shouting. The distance between speaker and listener is also an important factor. The setting of criteria must depend to some extent on the spectrum of the interfering background noise and how it relates to the speech spectrum. Low-frequency background sound is more acceptable than high-frequency sound except that low-frequency (500 Hz) pure tones have a strong masking effect.

A widely accepted standard way of assessing the problem is to use Speech Interference Levels (SIL) to provide a measurement of the background noise. The SIL is the average sound pressure level (SPL) in the three octave bands centred at 500 Hz, 1 kHz and 2 kHz. In some countries this is calculated in a slightly different way. Figure 2.11 shows the maximum SIL-values which can be tolerated for speech communication at the level and distance indicated. The problem depends upon the speaker as well as the listener: in using these data it must be recognised that the average person will raise his voice in response to high background noise levels and this feature will become part of any assessment. The upper limiting SIL-values for the indicated voice levels are approximately as shown in table 2.6; above these values, the speaker will attempt to raise his or her voice:

Table 2.6 SIL values for various voice levels.

Voice level	Upper limit SIL
Normal	40
Raised	55
Very loud	72
Shouting	85

It is sometimes convenient (though less accurate) to use measured dBA-values instead of SIL-values. If this is the case, an approximate equivalent SIL-value is obtained from the equation:

$$SIL = dBA - 9 \tag{2.2}$$

All values derived from figure 2.11 should be reduced by 5 dB if the speaker is female.

Example

The SIL in a small printing shop is 60 dB. Assess the problem of speech communication.

Solution

Considering first the voice levels that speakers will use: the table above shows that nothing less than a raised voice is likely to be used. From figure

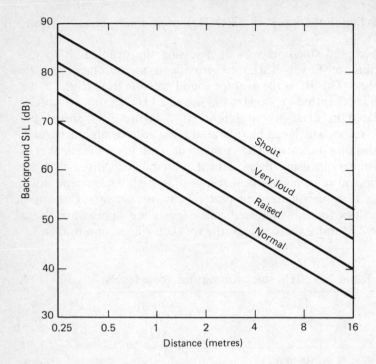

Figure 2.11   Maximum allowable background SIL to ensure speech communication at the voice levels shown

2.11 it follows that all communication at distances less than 1.5 m will be conducted with raised voices. Between 1.5 m and 3 m, a very loud voice will be required to ensure adequate communication. At distances between 3 m and 6 m, shouting will be required. Beyond a range of 6 m, even if the speaker shouts, intelligibility of communication will decrease.

REFERENCES

[1] International Standards Organisation, *ISO 532 Acoustics – method for calculating loudness levels* (1975).
[2] *The use of L_{eq} as an aircraft noise index.* DORA report 9023 (1990). Available from the CAA, 37 Gratton Road, Cheltenham, UK.
[3] J. W. Little and J. E. Mabry, Empirical comparisons of calculation procedures for estimating annoyance of aircraft flyovers, *Journal of Sound & Vibration*, Vol. 10, No. 1 (1969).
[4] J. R. Hassall and K. Zaveri, *Acoustic Noise Measurements*, Bruel & Kjaer (1988).
[5] *Noise at Work Regulations*, Health and Safety Executive, 1 Chepstow Place, London W2 4TF (1990).

3. Sound in three dimensions

I do loathe explanations.
J. M. Barrie, *My Lady Nicotine*

3.1 INTRODUCTION

The analysis in chapter 1 relates strictly to one-dimensional acoustic waves. This is either sound constrained in a tube of small diameter (compared with wavelength) or a plane-fronted wave in a three-dimensional open space. The simplicity of the one-dimensional analysis assists with an understanding of many of the important features of sound propagation. However, there are some features of importance which cannot be described or understood without an adequate consideration of sound in three dimensions. An obvious example is the directional behaviour of sound radiated from some types of source. In this chapter, the propagation of sound from a symmetric spherical source will first be considered. This will lead to an understanding of the efficiency of radiation of 3-D sources and the concepts of near and far fields. This analysis will be followed by a consideration of the directional radiation and reception properties of an oscillating flat surface such as a piston model loudspeaker or a disc microphone. Directional properties may only be understood by reference to this type of radiator or receiver because, by definition, a spherically symmetric radiator, which may be envisaged as a pulsating balloon, propagates sound equally in all directions. From these analyses the great importance of the ratio of size to wavelength may be appreciated, both as regards radiation efficiency and directionality.

3.2 SPHERICAL WAVES

The acoustic wave equation in one dimension has been shown in chapter 1 to be

$$\frac{\partial^2 u}{\partial t^2} - c^2 \frac{\partial^2 u}{\partial x^2} = 0 \tag{3.1}$$

or alternatively, in terms of pressure

$$\frac{\partial^2 p}{\partial t^2} - c^2 \frac{\partial^2 p}{\partial x^2} = 0 \tag{3.2}$$

It can also be shown, though there is no need here to go into details, that the equivalent equation in three dimensions is

$$\frac{\partial^2 p}{\partial t^2} - c^2 \nabla^2 p = 0 \tag{3.3}$$

where the Laplacian operator ∇^2 is defined in cartesian coordinates by the equation

$$\nabla^2 = \frac{\partial^2}{\partial x^2} + \frac{\partial^2}{\partial y^2} + \frac{\partial^2}{\partial z^2} \tag{3.4}$$

It has alternative forms in other coordinate sets. In spherical coordinates, as defined in figure 3.1, it is given by

$$\nabla^2 = \frac{1}{r^2} \frac{\partial}{\partial r} \left(r^2 \frac{\partial}{\partial r} \right) + \frac{1}{r^2 \sin\theta} \cdot \frac{\partial}{\partial \theta} \left(\sin\theta \frac{\partial}{\partial \theta} \right) + \frac{1}{r^2 \sin^2\theta} \cdot \frac{\partial^2}{\partial \phi^2} \tag{3.5}$$

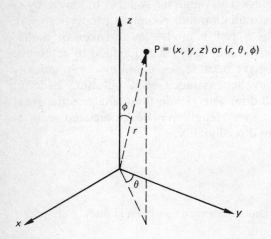

Figure 3.1

This is particularly relevant for a spherical source (which can be thought of as a pulsating balloon) and for a spherically symmetric field beyond the source for which

$$\frac{\partial p}{\partial \theta} = \frac{\partial p}{\partial \Phi} = 0$$

In this case the wave equation reduces to

$$\frac{1}{r^2}\frac{\partial}{\partial r}\left(r^2\frac{\partial p}{\partial r}\right) - \frac{1}{c^2}\frac{\partial^2 p}{\partial t^2} = 0 \tag{3.6}$$

This has the general solution for an outgoing wave of

$$p = \frac{A}{r}f\,(r - ct) \tag{3.7}$$

and this can be verified by substitution. The wavespeed c is still given by the formula in chapter 1, equation 1.13. For a harmonic wave

$$p = \frac{A}{r}\,e^{i(\omega t - kr)} \tag{3.8}$$

where A is a constant and k, the wave number, is given by

$$k = \frac{\omega}{c} = \frac{2\pi}{\lambda} \tag{3.9}$$

This equation shows that the magnitude of p is inversely proportional to r and, if extended to intensity (equation 1.16 in chapter 1), shows that the inverse square law must hold for energy propagation. This is to be expected from the simpler consideration of energy being spread outwards over spherical surfaces of increasing area.

At this point, a value for the impedance in a spherical wave can be derived. Considering the forces on a truncated conal element taken from a spherical annulus or shell of thickness δr, see figure 3.2, enables the force equilibrium equation in the radial direction to be written

$$\rho A \delta r \frac{\partial^2 u}{\partial t^2} = pA - \left(p + \frac{\partial p}{\partial r}\,\delta r\right)(A + \delta A) + \left(p + \frac{1}{2}\frac{\partial p}{\partial r}\,\delta r\right)\delta A$$

The derivation of the last term in this equation requires a little thought since the mean pressure on the side of the conal element is inclined at a

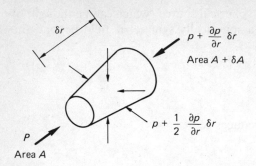

Figure 3.2

shallow angle to the transverse direction. If second-order small quantities are neglected this equation reduces to

$$\rho \frac{\partial^2 u}{\partial t^2} = -\frac{\partial p}{\partial r} \qquad (3.10)$$

In order to find the impedance in the wave we must find a relationship between pressure and particle velocity because, as in chapter 1, impedance is

$$Z = \frac{\textbf{acoustic pressure}}{\textbf{particle velocity}}$$

This can be achieved by integrating equation 3.10, having made a substitution for *p* from equation 3.8. First the substitution gives

$$\frac{\partial p}{\partial r} = -\frac{A}{r^2} \cdot e^{i(\omega t - kr)} + \frac{A}{r}(-ik)e^{i(\omega t - kr)} = -p\left(\frac{1}{r} + ik\right)$$

Thus

$$\frac{\partial^2 u}{\partial t^2} = \frac{p}{\rho}\left(\frac{1}{r} + ik\right)$$

Then integrating with respect to time, the particle velocity

$$\frac{\partial u}{\partial t} = \frac{p}{i\omega\rho}\left(\frac{1}{r} + ik\right)$$

Thus the specific impedance Z is given by

$$Z = \frac{p}{\partial u/\partial t} = \left(\frac{i\omega\rho}{\frac{1}{r} + ik}\right) = \frac{\rho ckr\ (i + kr)}{(1 + k^2r^2)} \tag{3.11}$$

If the impedance Z is expressed as

$$Z = R + iX$$

then

$$R = \rho c\ \frac{k^2r^2}{(1 + k^2r^2)} \tag{3.12}$$

and

$$X = \rho c\ \frac{kr}{(1 + k^2r^2)}$$

As in the one-dimensional analysis given in chapter 1, these equations apply equally well in the acoustic field and at the radiator surface. To help with the interpretation of these results, equation 3.9 is recalled so that the factor kr can be described as the non-dimensional ratio of spherical perimeter to wavelength.

Extreme cases

(a) If kr is very large, that is, r is large compared with wavelength, then

$$R = \rho c$$
$$X = 0$$

In the acoustic field this represents the condition a long way from the source where the wave is effectively flat-fronted. It therefore coincides with the previous result in chapter 1 for one-dimensional waves.

(b) If kr is very small then

$$R = \rho c(kr)^2$$
$$X = \rho c(kr)$$

Both components are now small compared with ρc, and R is smaller than X.

The detailed variations of R and X are shown in figure 3.3. The electric circuit analogy, introduced in chapter 1, can still be used. There is now a capacitive component (which can store energy but not dissipate it)

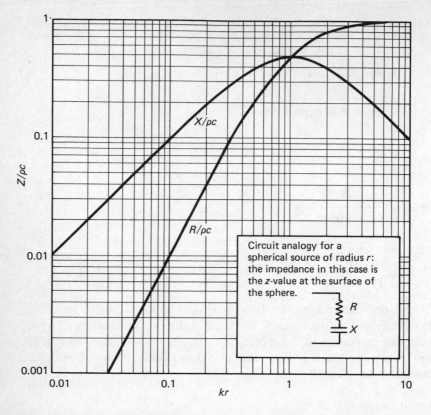

Figure 3.3   *The real and imaginary parts of the normalised specific acoustic impedance of a spherically symmetric field (equation 3.12)*

corresponding to *X* as well as the resistive component *R*, as for 1-D waves, which can dissipate energy.

It is most important to recognise that these results for impedance apply both in the free space surrounding a spherical source and also at its surface, where it may be considered as a driving point impedance.

Considering first the waves in free space, it has already been seen that at a considerable distance from a spherical source (*kr* very large) the waves are effectively plane and one dimensional. At medium distances (that is, when *kr* is between 0.1 and 10, say) there will be two components of the sound field; the first is associated with *R* and represents radiating energy and extends to infinity. This is called the 'far field'. The second is associated with *X* and radiates no energy. It extends only a short distance from the source and is therefore called the 'near field'. Thus, the nature of energy transfers in the wave fits with the explanation of the circuit analogy given above. Indeed, the circuit analogy, modified to include the capacitive element, is an accurate model of the three-dimensional acoustic system.

Next, consider the impedance value at the surface of a spherical source. If the source is large ($kr \gg 1$) it will have a driving point impedance which is almost entirely resistive because the wavelength is much smaller than the size of the source. The resistive value will also be close to $R = \rho c$ which is the greatest value that it can take. Thus the source is as 'efficient' as possible. On the other hand, if the wavelength is larger than the source then at the surface kr is small, so is R, and hence the sphere is a poor radiator of sound. An additional point is that all small sources, whether spherical or not, can be regarded as equivalent spheres, as will be demonstrated when directionality is considered later in this chapter. These conclusions, relating to source size, apply to all real three-dimensional sound sources, and this is illustrated in the problem below:

Example

A 300 mm (diameter) loudspeaker mounted in a small baffle box is radiating sound at 50 Hz. Calculate the extent of the near-field, the limit being defined as the surface where X/R has fallen to 1/10. What is the radiating efficiency of the loudspeaker at this frequency?

Solution

At this low frequency, where the wavelength is much larger than the source size, the loudspeaker may be assumed to behave like a spherical source (at higher frequencies the speaker behaves more like an oscillating piston and becomes more directional, as will be seen later in this chapter). Thus, from equation 3.12:

$$\frac{X}{R} \approx \frac{1}{kr} = \frac{\lambda}{2\pi r}$$

$$\lambda = 340/50 = 6.8 \text{ m}$$

Substituting for $X/R = 0.1$:

$$r = 10.8 \text{ m}$$

Thus it can be deduced that in an ordinary living room a listener is in the near-field of such a speaker at frequencies up to around 100 Hz or even higher.

Efficiency

There are only two qualities in the world –
efficiency and inefficiency.
G. B. Shaw

An ideal transducer will produce plane-wave radiating quality, that is to say R will be equal to ρc. For driving point impedances lower than this, the pressure will be less than it is possible to get for a given velocity. If the speaker is approximated to a sphere of 300 mm diameter then at the surface $r = 0.15$ m and

$$kr = \frac{2\pi(0.15)}{6.8} = 0.139$$

Using equation 3.12 for R:

$$R = \rho c(0.0188)$$

Thus the speaker is only 1.88 per cent 'efficient' at producing the ideal pressure at 50 Hz. Show that the efficiency at 500 Hz is 66 per cent.

What is the relevance of all this to noise control? Perhaps the most important feature to emerge is that large sources (compared with wavelength) are efficient radiators. This is why a guitar string has to be connected to a sound box if it is to make a reasonably loud sound. On the other hand, if there is a vibrating component in a machine this can make an unwanted loud sound if the machine is firmly connected to a large casing, or even to a wall or floor. Thus, in noise control, large surfaces connected to vibrating components should be avoided or uncoupled. This is one of the reasons why motor vehicle engines are mounted on vibration isolators. If it were not so the noise inside the vehicle would be unbearable.

Example

The noise level at the ear of a machine tool operator is to be reduced to a level below 90 dBA in order to meet the requirements of Health and Safety legislation. An investigation of the machine reveals that a flat panel, 0.3 m square, vibrates in resonance at about 500 Hz. The measured average displacement amplitude is 0.1 mm and the whole surface of the panel vibrates in phase. The operator's ear is at a distance of 1.5 m from the panel.

To establish whether treatment of the panel would be worthwhile, estimate the dBA contribution, at the operator's ear, due to the vibration.

Base your estimate on radiation from a sphere of equivalent surface area. Check whether it is reasonable to make the equivalent sphere assumption. Take into account the efficiency of radiation for the equivalent sphere using the data in figure 3.3. The British Standard A-weighting correction at 500 Hz is −3.5 dB.

Solution

The area of the panel is 0.09 m^2. An equivalent sphere of radius r is of surface area:

$$A = 4\pi r^2$$

Hence

$$r = \mathbf{0.085} \text{ m}$$

At a frequency of 500 Hz the wavelength is

$$\lambda = 340/500 = 0.68 \text{ m}$$

Thus the factor kr at the sphere surface is

$$kr = 2\pi r/\lambda = 2\pi(0.085)/0.68 = 0.78$$

This factor is less than unity so the equivalent sphere assumption is valid. From figure 3.3 the value of the radiation impedance at the sphere surface is

$$R = \mathbf{0.37}\rho c = 0.37 \times 407 = 150.6$$

Thus, the energy radiated, using the circuit analogy, is

$$U = \text{Area} \times (I^2 R) = \text{Area} \times (\text{velocity}^2 \times R)$$

If the surface displacement amplitude is 0.1 mm then, at 500 Hz, the velocity amplitude is

$$\frac{0.1 \times (500 \times 2\pi)}{1000} = 0.314 \text{ m/s}$$

The energy is thus

$$U = 0.09 \times (0.314^2 \times 150.6) = 1.34 \text{ Watts}$$

At the listener's ear (at a radius of 1.5 m) the intensity will be

$$I = 1.34/(4\pi \times 1.5^2) = 0.047 \text{ W/m}^2$$

It follows that the SPL in dB is

$$10 \log_{10}(0.047/10^{-12}) = 106.7 \text{ dB}$$

Making the A-weighting correction:

$$\text{dBA} = 103.2$$

This level is considerably in excess of the permitted value of 90 dBA, so treatment of this panel would be essential.

3.3   DIRECTIONALITY

> *If we have any kind of efficiency, very*
> *much of it is due to our narrowness.*
> P. G. Homerton, *The Intellectual Life*

In this section, we first have a qualitative description of the directional nature of sound radiation from a flat circular piston. This is followed by the corresponding description for the reception of sound by a similar flat circular microphone. The first corresponds roughly to the quality of radiation that is obtained from an ordinary loudspeaker. The second corresponds to reception properties for a flat circular membrane microphone. In fact, if the diameter of the two were the same, their directional properties would be identical. This is an example of the general reciprocal theorem for linear systems. It is therefore only necessary to look closely at the theory for either one or the other; either for the radiator or for the receiver. In the next few paragraphs, these qualitative descriptions are followed by the simple theory for a rectangular membrane microphone. This is chosen, in preference to the circular membrane, because the theory for the circular microphone involves the complication of Bessel functions. The nature of the results for the two are, in any case, very similar.

Figure 3.4 shows a flat piston oscillating sinusoidally in the direction indicated by the double arrow. The directional nature of radiation, when the diameter of the piston is large compared with the wavelength, can best be appreciated by considering the radiation from two areas X and Y on the surface of the piston which are half a wavelength apart, as shown. A listener at A directly in front of the piston will receive sound from the two areas in-phase because the path lengths from A to X and from A to Y are virtually identical. In fact, this will be true for all of the elemental areas

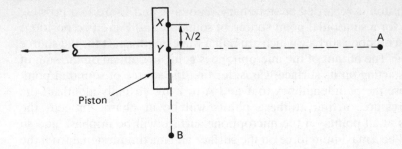

Figure 3.4

which add up to make the piston area as a whole. On the other hand, a listener at B, to the side of the piston, will receive sound from the two areas *X* and *Y* out of phase because the path length B to *X* is half a wavelength longer than the path length B to *Y*. Thus there will be cancellation. If the area of the piston as a whole is considered, there can be a large degree of cancellation at B. Thus, the sound level at B can be much less than that at A. There can, in fact, be other directions where cancellation is complete. The argument obviously depends strongly on the fact that the size of the piston is large compared with the wavelength. If, on the other hand, the size is less than the wavelength, similar arguments will show that radiation is almost uniform in all directions. Figure 3.5 shows how these directional properties may be summarised by means of polar plots of intensity. It is seen that the factor *kr*, which in this case is the ratio of piston perimeter to wavelength, is most important.

Next, figure 3.4 will again be used to describe the directional nature of sound reception by a circular piston microphone. In this description, the

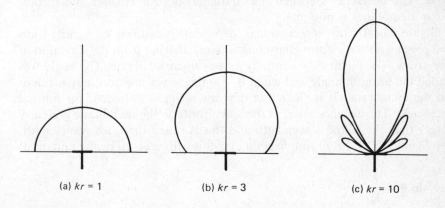

(a) *kr* = 1   (b) *kr* = 3   (c) *kr* = 10

Figure 3.5

circular piston is regarded as stationary. Points A and B are two possible positions for a sinusoidal point source of sound. *X* and *Y* are two positions on the microphone surface which are half a wavelength apart for the source frequency. The output of the microphone is going to depend on the sum of pressures acting on its surface. Consider first the source of sound at point A: because the path lengths A to *X* and A to *Y* are virtually identical, the sound pressures acting at these points will be in phase. Indeed, the pressures at all points on the microphone surface will be in phase and so there will be a maximum force on the surface and maximum signal from the microphone. Now consider the source of sound at B: the path lengths B to *X* and B to *Y* are half a wavelength different. Thus when the acoustic pressure is a positive maximum at *X* it will be a negative minimum at *Y* and vice versa. In the summation of pressures to find the force on the microphone membrane, the effect at these two areas *X* and *Y* will always cancel out. This argument is closely analogous to that for the piston radiator, and identical polar plots of reception sensitivity to those shown in figure 3.5 will be obtained.

3.4 MEMBRANE MICROPHONES: A MORE QUANTITATIVE THEORY

Membrane microphones usually have a flat circular membrane which is tensioned and which is the sensitive element (see chapter 6 for more general information about microphones). The deflection of the membrane is proportional to the force acting on it and this is the sum or integral of the pressure acting over its surface. In a well-designed microphone, the electrical output is proportional to force. In this section, the theory is described for a rectangular membrane with waves arriving parallel to one edge. The difference between the characteristics for circular and rectangular membranes is minimal.

Figure 3.6(a) shows a rectangular membrane measuring *L* by unity (into the page) with wavefronts (harmonic waves) arriving from the direction of the arrow. The picture is essentially a two-dimensional one. The angle θ is called the azimuth angle and when $\theta = 0$ the waves are normally incident on the membrane. It is clear that maximum signal will occur for normal incidence. The distance apart of the wavefronts on the membrane is clearly larger than the sound wavelength and this is called the trace wavelength, see figure 3.6(b). From simple geometry this will be given by the equation:

$$\sin \theta = \frac{\lambda}{\lambda_t}$$

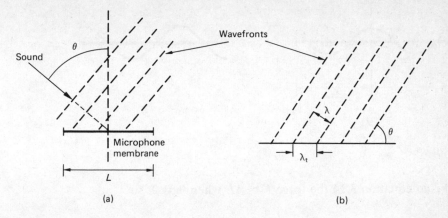

Figure 3.6

or

$$\lambda_t = \lambda/\sin \theta \tag{3.13}$$

The maximum force on the membrane occurs when the pressure pattern is symmetrical about the middle, $x = 0$, as in figure 3.7(a). It is zero a quarter of a period earlier or later when there is complete cancellation, as in figure 3.7(b). Clearly the force will vary sinusoidally with time, the peak amplitude being given by the integral of the pressure in figure 3.7(a).

For this diagram:

$$p = A\cos\left(\frac{2\pi x}{\lambda_t}\right)$$

where A is an arbitrary pressure amplitude in the wave. The total force on the membrane will then be

$$F = \int_{-\frac{L}{2}}^{\frac{L}{2}} p \, \mathrm{d}x = \frac{A\lambda_t}{\pi} \sin\left(\frac{\pi L}{\lambda_t}\right)$$

Substituting from equation 3.13:

$$F = \frac{A\lambda}{\pi\sin \theta} \sin\left(\frac{\pi L}{\lambda} \sin \theta\right) \tag{3.14}$$

The output signal is proportional to this quantity. It can be normalised by making the sensitivity to normally incident sound ($\theta = 0$) equal to unity.

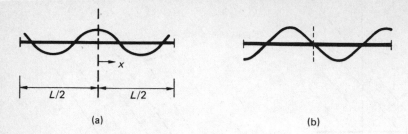

(a)   (b)

Figure 3.7

From equation 3.14 the force $F = AL$ when $\theta = 0$, so

$$\text{Normalised sensitivity} = N = \frac{\sin\left\{\left(\dfrac{\pi L}{\lambda}\right)\sin\theta\right\}}{\left\{\left(\dfrac{\pi L}{\lambda}\right)\sin\theta\right\}} \tag{3.15}$$

N is the $\sin(x)/x$ function seen in figure 3.8, where x depends on the azimuth angle θ and the size to wavelength ratio L/λ.

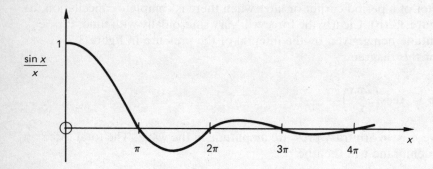

Figure 3.8

The first zero is at $x = \pi$ or when

$$\sin\theta = 1/(L/\lambda)$$

Thus if L/λ is great, θ_1 for first zero is small and a polar plot of this is shown in figure 3.9. This figure is closely similar to the polar plot shown in figure 3.5. The major lobe is relatively narrow and the beamwidth for reception is $2\theta_1$. In these circumstances the microphone is quite directional.

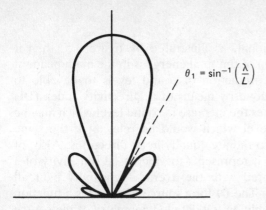

$$\theta_1 = \sin^{-1}\left(\frac{\lambda}{L}\right)$$

Figure 3.9

Example: calculation for a one-inch microphone (such as the B & K type 4145)

Directionality becomes important when $\lambda/L < 1$; in this case when λ is smaller than 2.54 cm. The corresponding critical frequency is about 13.5 kHz. Thus the microphone becomes directional towards the end of the audio range (0–16 kHz). Generally, it is required that microphones be omni-directional in the audio range and this is why the one-inch microphone is the largest normally used. Occasionally there is demand for a very directional microphone. This may be achieved, in principle, by using membrane microphones of large surface area. However, this usually leads to problems of either mechanical resonance of the membrane or high cost. An alternative is to employ a small microphone combined with a parabolic reflector. The effective size of the microphone is then the size of the reflector.

Peak sensitivity in the first side lobe

This occurs at $x = 3\pi/2$ (see figure 3.8); At this point $N = 2/3\pi$ and in dB terms the size of this peak is

$$20 \log_{10}(2/3\pi) = -13.5 \text{ dB}$$

This figure is, of course, relative to 0 dB for normal incidence. The side lobes are thus relatively weak.

3.5   DIRECTIVITY

If a source of sound is directional, as illustrated by figure 3.5, then it follows that a greater proportion of the total energy is in the main beam of sound. It is important in the prediction of sound levels to be able to quantify this increase. This is done by means of a Directivity Index (**DI**) measured in dB. The **DI** describes the increase in sound level which may be expected compared with the level which would be relevant if the same amount of sound energy were to radiate equally in all directions. Also, of course, because of reciprocity, it represents the increased sensitivity of a directional microphone compared with the average sensitivity over all possible directions of reception. The **DI** for a source of sound is a function of two factors. It depends not only on the detailed geometry of the source but also on the local presence of reflecting surfaces. Approximate calculations for radiation from most flat surfaces can be based on the **DI**-values for a flat circular piston and these are shown in figure 3.10.

However, caution should be adopted if the surface is either long and thin rather than equi-axed, or if the surface is curved. It may be necessary to seek more detailed data in such cases. The effect of source proximity to reflecting surfaces is summarised for simple surfaces in table 3.1. The effects of the two factors on **DI** are not necessarily additive because the main beam of sound may not be pointed at reflecting surfaces. A measure of judgment is required in combining the two factors if both are present. It is perhaps obvious that, for highly directional sources, even without

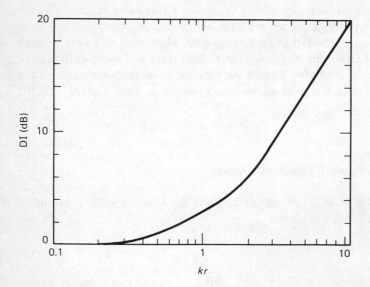

Figure 3.10

reflecting surfaces, the orientation of the source of sound is very important.

Table 3.1 Directivity Index (*DI*) for a spherical source as a function of nearby surfaces

Source/surface type	DI
Source in free space	0 dB
Source close to a flat surface of large extent (for example, near the surface of open ground)	3 dB
Source close to the intersection of two flat surfaces of large extent (for example, near the ground and a wall)	6 dB
Source near the ground and near a wall or other surface of limited extent	< 6 dB > 3 dB
Source close to the intersection of three flat surfaces of large extent (for example, in the corner of a room)	9 dB

3.6 SCREENS

Screens are used in noise control to attenuate unwanted sounds by placing a physical obstacle between the source and observer. Screening does not provide perfect attenuation because of diffraction around and transmission through the screen. The latter mechanism, transmission, is usually insignificant for solid screens. The sound observed at any point in the field is composed of two parts: (a) direct sound, and (b) sound diffracted around the top of the screen.

Referring to figure 3.11, if the observer is at O then the first of these two parts is zero. The second (diffracted) part depends on the ratio of the effective height *H* to the wavelength of the sound. The greater the height in wavelengths, the greater will be the sound attenuation. Hence, in these circumstances the perceived sound at O is weak and the screen is doing its job well. If, however, the screen is small compared with the wavelength, it will then be relatively ineffective. These general considerations lead to the information given on figure 3.11(b) for such screens. This information may be used to make calculations of the effectiveness of such screens provided, of course, that direct transmission can be neglected. This condition is not

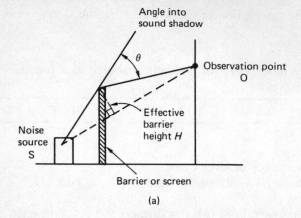

(a)

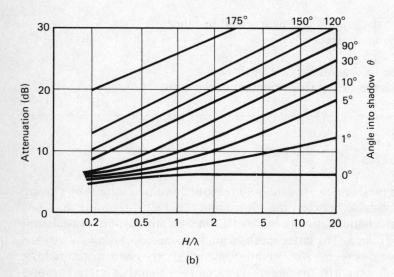

(b)

Figure 3.11   Shielding provided by barriers such as solid screens or walls

true for light fabric materials (hessian sacking, for example) or for belts of trees or shrubs, and generally they do not make good sound screens. Table 3.2 is a set of data useful for interpreting the effect of figure 3.11(b) for some common standard effective heights.

Table 3.2 Data for interpreting figure 3.11(b)

| f | λ | H/λ for | | | |
		$H = 0.5$	$H = 1.0$	$H = 1.5$	$H = 2.0$
62.5	5.45	0.092	0.184	0.28	0.37
125	2.72	0.184	0.37	0.55	0.74
250	1.36	0.37	0.74	1.10	1.47
500	0.68	0.74	1.47	2.2	2.9
1000	0.34	1.47	2.9	4.4	5.9
2000	0.17	2.9	5.9	8.8	11.7
4000	0.085	5.9	11.7	17.6	23.3
8000	0.043	11.7	23.3	35	47

Example

The A-weighted mean octave-band spectrum of traffic noise measured adjacent to a house situated 100 m from a road is:

f (Hz)	dBA
125	52
250	60
500	65
1000	64
2000	61
4000	58

To reduce this noise it is proposed to erect a wall, 2 m high, at the roadside. Use the data shown on figure 3.11 to determine the reduction which may be achieved in the overall dBA noise value. For the purposes of the calculation, assume that the traffic is immediately adjacent to the wall and 0.5 m above the road.

Solution

The overall dBA-value before erecting the wall is calculated using the nomogram from chapter 1, see figure 1.7. Using the figures in the last column of the table above, this sum is 69.4 dBA. The effect of the wall may now be calculated in tabular fashion from figure 3.11 using the curve for $\theta = 90°$. The effective barrier height H is 1.5 m. The table below shows height in wavelengths.

f (Hz)	Wavelength (m)	$\dfrac{H}{\lambda}$	Attenuation (dB)	New dBA
125	2.72	0.55	13	39
250	1.36	1.10	16	44
500	0.68	2.21	18	47
1000	0.34	4.41	21	43
2000	0.17	8.82	24	37
4000	0.085	17.6	27	31

The sum of figures in the final column, obtained as before, is 50.4 dBA. The reduction in the overall dBA-value is thus 19 dB.

Further example

Consider the acoustic design of the violin (or any other member of this family of musical instruments). As far as radiation is concerned, what are the objectives and how are they achieved? The mid-range of the violin is the A (440 Hz) above middle C and the sound box is about 30 cm long.

Solution

There are two requirements for the sound radiation from musical instruments. First, the instrument should be a reasonably efficient radiator if it is to produce sound with a minimal expenditure of energy. Secondly, the sound output should not be highly directional, particularly so for the violin where the musician can be expected to move about (in an angular sense) during his performance. Referring to table 3.3, it can be seen that these two requirements are in conflict. The compromise which is achieved in the evolution of the instrument is to make

$$kr = 1$$

This can be demonstrated as follows: for the mid-range note

$$f = 440 \text{ Hz}; \lambda = 340/440 = 0.773 \text{ m}$$

The perimeter of the sound box for the violin is about 80 cm. Hence, the ratio of size to wavelength is

$$kr = \text{perimeter/wavelength} = 0.8/0.773 \approx 1$$

3.7 SOUND IN THE OPEN AIR

There are three principal effects in the open air which usually tend to reduce the perceived sound intensity compared with that calculated from the simple inverse square law. These are:

(a) sound absorption in the air itself;
(b) refraction of sound away from the ground caused by wind gradients;
(c) refraction of sound away from the ground caused by temperature gradients.

Although these usually cause the sound intensity to be reduced, the latter two can also, in some unusual circumstances, cause the sound intensity to be raised.

The absorption of sound energy in the air is, for most noise control purposes, rather insignificant. The absorption figure A is generally quoted in dB per 100 m and the classical theory gives the formula

$$A = Bf^2 \tag{3.16}$$

where the constant B has a value of about 40 million when the frequency f is expressed in Hz. However, the absorption is strongly dependent on humidity in a non-linear way and this is shown in figure 3.12, together with the classical line given by equation 3.16. If values are required they can be estimated by interpolation.

There can also be absorption at ground level caused by the nature of the ground surface. Hard surfaces provide virtually no such absorption but natural vegetation provides between 2 and 3 dB per 100 m at frequencies below 1000 Hz.

Refraction may be described as bending of the line of propagation of sound caused by changes in the speed of sound. It is closely analogous to the bending of light when it passes from one medium to another, as in a prism or lens. Under windy conditions, close to the ground, there is a vertical wind speed gradient in the boundary layer, with low velocity close to the ground and higher velocities at greater heights. Thus, a sound wave travelling upwind tends to be bent upwards while downwind it is bent downwards. At ground level there is, as a result, a concentration of energy downwind. It is difficult to quantify this effect because it becomes important only at relatively long range (of the order of 1 km or more) and there are considerable secondary effects caused by air turbulence and the nature of the terrain. However, to give some idea of orders of magnitude a typical increase downwind at a range of 1 km might be 10 dB. Upwind it is possible for there to be a 'shadow zone' where virtually no sound will

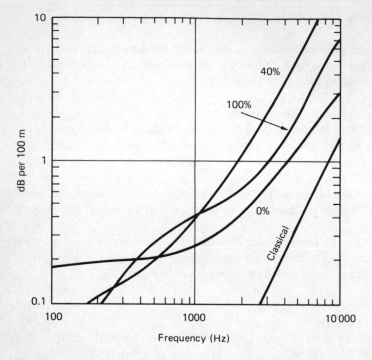

Figure 3.12   Absorption of sound in air as a function of frequency for various relative humidity values

penetrate. When there is a prevailing wind this is important in positioning noise sources, such as pop festivals, motorways or industrial estates, in relation to a town.

Refraction due to thermal gradients in the atmosphere results from the temperature dependence of the speed of sound (equation 1.9).In a normal atmosphere, temperature decreases with height. As a result, sound from a source on the ground is bent upwards and so a listener at ground level, some distance away, is subjected to a lower acoustic intensity than would be predicted by the inverse square law. This is, again, difficult to quantify though a figure of 10 dB extra attenuation at a range of 1 km may be taken as a rough guide. Under these conditions, a 'shadow zone' can also form at longer range. Unusually, a temperature inversion can occur, often in summer and often in the evening or early morning. When this happens, the sound is bent downwards and large increases can occur at longer ranges, say 3 to 10 km. There can be, at the same time, a zone of silence at shorter range. This is the acoustic equivalent of a mirage.

3.8   SUMMARY

- A piston radiator of small size compared with wavelength radiates more or less uniformly in all directions, much as the spherical source does. It may be inferred that when a source OF ANY SHAPE is small compared with wavelength, it radiates uniformly in all directions. The radiation strength is, in fact, proportional to the amplitude of the volume displacement at the source. In these circumstances, the results from the spherical theory can be used.
- Out-of-doors directionality can be used as a noise control measure, that is, by turning the source to the optimum orientation. This cannot normally be done indoors because multiple reflections from the walls ensure a more or less non-directional radiation, whatever the source. Predictions of these effects are difficult and often inaccurate.
- Table 3.3 shows a summary, in brief, of the quality of sound sources of various sizes as described in this chapter.

Table 3.3 General properties of sound sources of different sizes (see figures 3.3 and 3.5 for details).

Property	Small source $kr < 1$	Large source $kr > 1$
Radiation efficiency	Low	High
Directional radiation	Weakly or not at all	Strongly

$kr = 2\lambda r/\lambda = $ 'size'/wavelength.

4. Analysis of acoustic and vibration signals

MORIARTY: How are you at mathematics?
SECOMBE: I speak it like a native!
Goon Show: The Affair of the Lone Banana

4.1 INTRODUCTION

In previous chapters we have seen how sound waves propagate as small pressure disturbances in a gas or liquid, or as changes in stress in a solid. Before we can study sound waves we need to be able to measure them. This is achieved by using a sensor (such as a microphone) which converts the pressure or stress change to an electrical form. Once we have made an electrical analogue of the sound wave we can apply the analysis techniques described in this chapter to the signal. The same applies to vibration signals which require analysis.

The aim of this chapter is to introduce the reader to frequency domain analysis, and in particular to the use of the Fourier Transform (the FT) to examine data in both time and frequency domains. The topic is one of some complexity and no attempt at mathematical rigour has been made. The intention is simply to give the reader a 'feel' for the concepts involved.

The record of a signal's past behaviour is referred to as its time history. Observed time histories of signals representing physical phenomena can be classified as either deterministic or random. Deterministic signals may be described explicitly by a mathematical relationship, while random signals by their very nature cannot be predicted and have to be described in terms of probability statements and statistical averages.

Most of the acoustic phenomena discussed in this book are deterministic. This is because acoustic signals are commonly either periodic or transients, and because the response of a linear system to deterministic signals (and to some extent to random signals) may generally be described, using the principle of superposition, in terms of the sum of the individual responses to a series of harmonic signals.

A periodic signal repeats itself at a specified time interval T, such that

$$f(t + T) = f(t) \tag{4.1}$$

It can be either sinusoidal or complex periodic. A complex periodic signal consists of a sum of harmonic (that is, sine and cosine) terms which combine to form a complex waveform which repeats itself over a specified time interval. Such signals can be described by a Fourier series. The frequencies of the higher terms in such a series are always integral multiples of the fundamental frequency (which is the reciprocal of the repetition period T).

Non-periodic deterministic signals can also be described by explicit mathematical relationships, but do not have a functional form which exactly repeats itself over a specified time interval as indicated by equation 4.1. These signals can be either transient or almost periodic. Almost periodic signals arise when two or more harmonic signals with arbitrary frequencies are added together. The higher frequency terms in the resulting sum are not generally integral multiples of the lowest frequency term. Transient signals include all non-periodic signals other than almost periodic signals. Examples include damped sinusoids, exponentially rising and falling signals, and square pulses. Non-periodic signals can be described by a Fourier Transform.

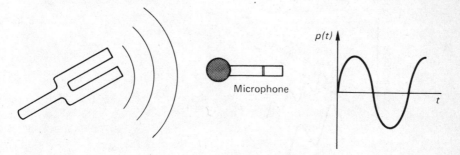

Figure 4.1   The mechanical vibration of a tuning fork causes sound waves

The traditional way of observing signals is to examine them in the time domain. As we saw earlier, the time domain consists of records of the behaviour of a system as time passes. For instance, figure 4.1 shows a tuning fork together with a record of the pressure disturbance it produces after being struck. The sensor used to convert the pressure changes into electrical changes is usually a microphone.

The French mathematician Fourier showed that any periodic waveform can be represented as the sum of a number of sinusoids. By selecting the

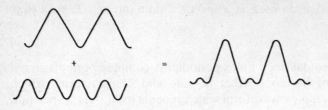

Figure 4.2

right amplitudes, frequencies and phases, any periodic waveform may be synthesised. For example, the waveform of figure 4.2 is produced by the addition of a pair of sinusoids.

Figure 4.2 shows the addition of two frequencies in the time domain. A different way of looking at this addition is to view it as it would be presented on the screen of a spectrum analyser. It is important to see that there is no conflict between these alternative representations of the situation. We are simply considering the same piece of information from a different viewpoint. A simple analogy is shown in figure 4.3.

If the three-dimensional graph is viewed by looking along the frequency axis, the representation of 4.3(b) is obtained when the waveforms are

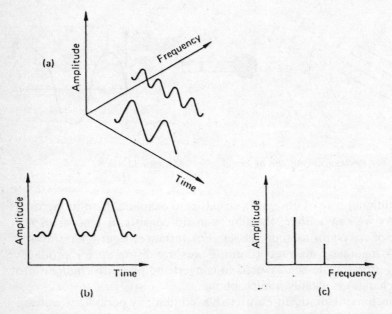

Figure 4.3   (a) Time and frequency domains represented in three dimensions; (b) time domain view; (c) frequency domain view

added. This is the time domain view of the waveform which we had in figure 4.2, obtained by adding together the component sinusoids at each moment in time. The axes are amplitude and time.

However, if we look along the time axis as in figure 4.3(c), a different view appears. The axes are now amplitude and frequency (time disappears since it is out of the page). The signal is observed in the frequency domain. Each sinusoidal component of the complex waveform appears as a vertical line. The height of each line represents the amplitude of that component, and the line's position indicates its frequency. Since each line represents a sinusoid, this representation characterises the waveform in the frequency domain. This representation is known as the amplitude spectrum of the waveform, and (apart from the phase of each component, which will be discussed later), it contains all the information necessary to reconstruct the complex waveform from its sinusoidal components.

Obviously, it would be unwise to push this geometrical analogy too far. For one thing, it neglects the phase of each component. Nevertheless, it provides a reasonable conceptual model of the difference between working in the time and frequency domains.

Since no extra information is provided by examining a signal in the frequency domain, and since the engineer requiring data in this form usually has to go to some lengths to get it, the reader may well wonder whether it is worth the trouble. The following examples will illustrate the utility of the method.

We might want to measure the level of distortion in the sound from a loudspeaker driven by an oscillator. The oscillator produces a pure sinusoid, but the sound from the loudspeaker will inevitably be distorted to some extent because of imperfections in the mechanical and electrical design.

Alternatively, suppose we want to monitor a bearing and detect the first signs of its failing in a noisy machine. The problem here is to detect a low-amplitude signal in the presence of a large amount of noise.

In each case we are trying to detect a low-amplitude signal in the presence of a second signal with a much larger amplitude at a different frequency. Figures 4.4(a) and (b) show what happens – the smaller signal is masked by the larger in the time domain, but both components of the signal appear clearly when it is viewed in the frequency domain (particularly if a logarithmic amplitude scale is used).

Let us consider the appearance of a few common waveforms in both the time and frequency domains. Figure 4.5 shows four signals (a sinusoid, a square wave, a damped oscillation and a sharp impulse) in both domains. Obviously, the amplitude spectrum of a pure sinusoid consists of a single line. A square wave on the other hand is made up from an infinite sum of harmonically related sinusoids. We should expect this, since a square wave is a complex periodic function which can be represented as a Fourier series.

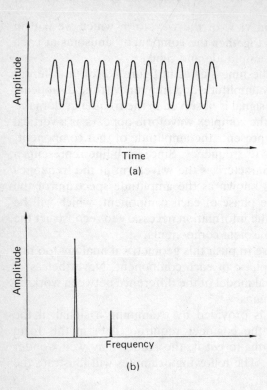

Figure 4.4 *(a) Small signal masked by larger in time domain; (b) both signals apparent in frequency domain*

Figures 4.5(a) and (b) illustrate the fact that periodic signals always have line spectra. Figure 4.5(c) on the other hand is a transient, and it can be seen that the corresponding spectrum is continuous rather than made up from discrete lines. We shall return to this point later when we consider the Fourier transform.

Figure 4.5(d) is a sharp impulse in the time domain, and appears as a broad band of energy in the frequency domain. This is because in general any function which is 'spiky' in one domain will appear spread out in the other. To mathematicians an impulse is a strictly defined waveform, often called a delta function, which has infinite amplitude, zero width and unit area. The frequency domain representation of a delta function contains equal energy at all frequencies, and appears as a horizontal line. Obviously such a waveform cannot exist in practice, but a sharp pulse such as that shown in figure 4.5(d) is a reasonable approximation and will contain energy over a wide range of frequencies.

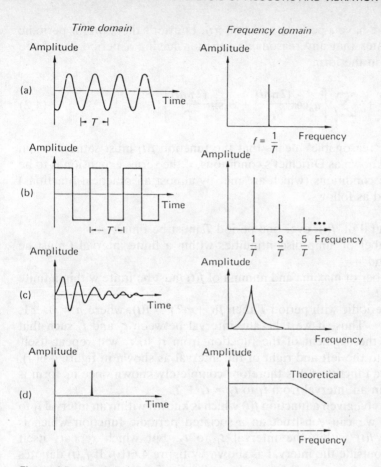

Figure 4.5   Signals in the time and frequency domains

4.2   FOURIER SERIES

It is well-known that a stretched string has a number of different natural frequencies. The lowest of these is called the fundamental frequency, f_1. The higher modes will have frequencies f_2, f_3 etc. and are called the harmonics or overtones. These frequencies can be excited simultaneously with the fundamental, and the resulting waveform (which is a sum of such sinusoids) will have a complex appearance. It will still however be periodic with a period $T = 1/f$. Thus, we see that the summation of a number of harmonically related waveforms gives rise to a complex periodic waveform. Conversely, any complex periodic waveform may be analysed and its constituents found as a set of harmonically related sinusoids.

Suppose we have a periodic function $f(t)$. Fourier's theorem for periodic functions states that any reasonable function having a period T may be represented in the form

$$f(t) = \frac{a_0}{2} + \sum_{n=1}^{\infty} \left\{ a_n\cos\frac{(2\pi nt)}{T} + b_n\sin\frac{(2\pi nt)}{T} \right\} \tag{4.2}$$

The term 'reasonable' means that the function $f(t)$ must satisfy certain conditions, known as Dirichlet's conditions, if the series expansion is to be valid. These conditions (which are met by almost all practical functions) can be stated as follows:

(a) the integral of $|f(t)|$ over one period T must be finite;
(b) the number of jump discontinuities within a finite interval t must be finite, and
(c) the number of maxima and minima of $f(t)$ must be finite within a finite interval t.

If $f(t)$ is periodic with period T then $f(t + n T) = f(t)$, where $n = 0, \pm 1, \pm 2, \pm 3 \ldots$. Thus, if we take any interval between t_1 and t_2 such that $t_2 = t_1 + T$, the segment of the function from t_1 to t_2 will repeat itself indefinitely to the left and right of the interval as shown in figure 4.6(a). Any periodic function $f(t)$ is therefore completely known once its form is known within an interval from t_1 to $t_2 = t_1 + T$.

Alternatively, given a function $f(t)$ which is known within an interval t_1 to $t_2 = t_1 + T$, we can construct an associated periodic function which is identical to $f(t)$ within the interval t_1 to t_2, but which repeats itself indefinitely outside the interval as shown by figure 4.6(b). If $f_p(t)$ denotes this associated periodic function, a formal definition of $f_p(t)$ is

$$f_p(t) = f(t) \qquad\qquad \text{where } t_1 < t < t_2$$
$$f_p(t + nT) = f(t) \qquad\qquad n = \pm 1, \pm 2, \pm 3, \ldots$$

To obtain expressions for the coefficients a_n and b_n in the Fourier series is straightforward. If equation 4.2 is integrated, we obtain

$$\int_{t_1}^{t_2} f(t)dt = \frac{a_0}{2}\int_{t_1}^{t_2} dt + \sum_{n=1}^{\infty}a_n\int_{t_1}^{t_2}\cos\frac{(2\pi nt)}{T}dt + \sum_{n=1}^{\infty}b_n\int_{t_1}^{t_2}\sin\frac{(2\pi nt)}{T}dt$$

and since

$$\int_{t_1}^{t_2}\cos\frac{(2\pi nt)}{T}dt = 0 \quad \text{and} \quad \int_{t_1}^{t_2}\sin\frac{(2\pi nt)}{T}dt = 0 \text{ for all } n$$

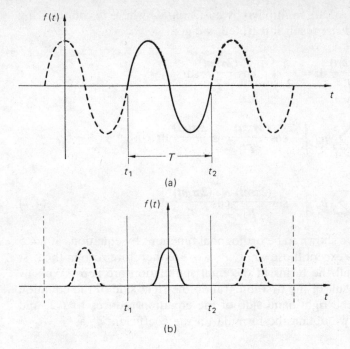

Figure 4.6 *Periodic and associated periodic functions*

the expression for a_0 follows immediately:

$$a_0 = \frac{2}{T} \int_{t_1}^{t_2} f(t)\mathrm{d}t \tag{4.3}$$

To obtain expressions for a_n and b_n we shall need to use some important properties of sine and cosine known as the orthogonality relationships. These are listed below:

$$\int_{t_1}^{t_2} \cos\frac{(2\pi nt)}{T} \cos\frac{(2\pi mt)}{T} \, \mathrm{d}t = 0 \qquad (m \neq n)$$
$$= T/2 \qquad (m = n) \tag{4.4}$$

$$\int_{t_1}^{t_2} \sin\frac{(2\pi nt)}{T} \sin\frac{(2\pi mt)}{T} \, \mathrm{d}t = 0 \qquad (m \neq n)$$
$$= T/2 \qquad (m = n) \tag{4.5}$$

$$\int_{t_1}^{t_2} \sin\frac{(2\pi nt)}{T} \cos\frac{(2\pi mt)}{T} \, \mathrm{d}t = 0 \qquad (m \neq n)$$
$$= 0 \qquad (m = n) \tag{4.6}$$

If both sides of 4.2 are multiplied by $\cos(2\pi mt/T)$ where m can take any integer value, and the result integrated, we get:

$$\int_{t_1}^{t_2} f(t)\cos\frac{(2\pi mt)}{T}\,dt = \frac{a_0}{2}\int_{t_1}^{t_2}\cos\frac{(2\pi mt)}{T}\,dt$$

$$+ \sum_{n=1}^{\infty} a_n \int_{t_1}^{t_2} \cos\frac{(2\pi nt)}{T}\cos\frac{(2\pi mt)}{T}\,dt$$

$$+ \sum_{n=1}^{\infty} b_n \int_{t_1}^{t_2} \sin\frac{(2\pi nt)}{T}\cos\frac{(2\pi mt)}{T}\,dt \qquad (4.7)$$

Sine and cosine are shown to be orthogonal functions by equations 4.4, 4.5 and 4.6. All terms except those where $n = m$ are therefore zero in the first summation, and all the terms in the final summation are zero. We saw earlier that the result of the first integral on the right side of the equation was zero. Thus, the right-hand side of the equation reduces to $T/2$, and replacing m by n we obtain the formula for the coefficient a_n as

$$a_n = \frac{2}{T}\int_{t_1}^{t_2} f(t)\cos\frac{(2\pi nt)}{T}\,dt \qquad (4.8)$$

Similarly, integration after multiplying by $\sin(2\pi mt/T)$ yields an expression for b_n:

$$b_n = \frac{2}{T}\int_{t_1}^{t_2} f(t)\sin\frac{(2\pi nt)}{T}\,dt \qquad (4.9)$$

To sum up, we have found expressions for the coefficients a_0, a_n and b_n and can thus express any periodic function $f(t)$ as a Fourier series by the use of equations 4.2, 4.3, 4.8 and 4.9. Even if $f(t)$ is not periodic, a Fourier series can still be used to represent $f(t)$ inside the interval t_1 to $t_2 = t_1 + T$. Outside the interval, the series will produce an associated periodic function.

There would be little point in representing a function such as $f(t)$ by a Fourier series if an infinite number of terms of the series had to be evaluated. Fortunately however, it can be shown that a Fourier series is highly convergent, and a good approximation to $f(t)$ is usually obtained by the sum of only a few terms.

A continuous, periodic signal always has a magnitude spectrum consisting of discrete lines. The height of the nth line of such a spectrum is given by $(a_n + b_n)^{1/2}$. Its phase is given by $\tan^{-1}(b_n/a_n)$.

4.3 EVEN AND ODD FUNCTIONS

If $f(t)$ is an even or odd function, the series expansion of $f(t)$ is simplified, with either the a_n or b_n coefficients becoming zero.

If $f(t)$ is an even function (that is $f(t) = f(-t)$) the integrand of equation 4.9 is odd, because it is a product of the even function $f(t)$ and the odd function $\sin(2\pi nt/T)$. Since the integral of an odd function over an interval symmetric about the origin is zero, the coefficients b_n are zero. An even function can therefore be represented by a Fourier cosine series:

$$f(t) = \frac{a_0}{2} + \sum_{n=1}^{\infty} a_n \cos\frac{(2\pi nt)}{T} \tag{4.10}$$

If $f(t)$ is odd, that is $f(t) = -f(-t)$, the integrands in expressions 4.3 and 4.8 for a_0 and a_n are both odd functions of t. Once again we are integrating an odd function over an interval which is symmetric about the origin, and a_0 and a_n are zero. An odd function may therefore be represented by a Fourier sine series:

$$f(t) = \sum_{n=1}^{\infty} b_n \sin\frac{(2\pi nt)}{T} \tag{4.11}$$

It is often useful to be able to predict analytically the frequency content of a periodic signal. This is relatively straightforward for most common waveforms, as shown by the following example.

Example

Figure 4.7 is a sketch of a triangular wave, which can be expressed as follows:

$$f(t) = 1 + \frac{4t}{T} \quad \text{for} \quad \frac{-T}{2} < t < 0$$

$$f(t) = 1 - \frac{4t}{T} \quad \text{for} \quad 0 < t < \frac{T}{2}$$

Since $f(t) = f(-t)$ the function is even, and the coefficients b_n will be 0. The coefficient a_0 is given by equation 4.3 as

$$a_0 = \frac{2}{T} \int_{t_1}^{t_2} f(t)dt = \frac{2}{T} \int_{-\frac{T}{2}}^{0} (1 + 4t/T)dt + \frac{2}{T} \int_{0}^{+\frac{T}{2}} (1 - 4t/T)dt = 0$$

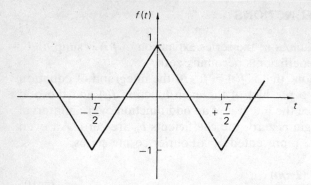

Figure 4.7

The coefficients a_n are found from equation 4.8:

$$a_n = \frac{2}{T} \int_{-T/2}^{+T/2} f(t) \cos\left(\frac{2\pi nt}{T}\right) dt$$

$$= \frac{2}{T} \int_{-T/2}^{+T/2} \cos\left(\frac{2\pi nt}{T}\right) dt + \frac{2}{T} \int_{-T/2}^{0} \frac{4t}{T} \cos\left(\frac{2\pi nt}{T}\right) dt$$

$$- \frac{2}{T} \int_{0}^{+T/2} \frac{4t}{T} \cos\left(\frac{2\pi nt}{T}\right) dt$$

$$= \frac{8}{T^2} \int_{T/2}^{0} (-t) \cos\left(\frac{2\pi nt}{T}\right) d(-t) - \frac{8}{T^2} \int_{0}^{T/2} t \cos\left(\frac{2\pi nt}{T}\right) dt$$

$$= -\frac{16}{T^2} \int_{0}^{T/2} t \cos\left(\frac{2\pi nt}{T}\right) dt$$

$$= -\frac{16}{T^2}\left\{\left[t\,\frac{T}{2\pi n}\sin\left(\frac{2\pi nt}{T}\right)\right]_{0}^{T/2} - \frac{T}{2\pi n} \int_{0}^{T/2} \sin\left(\frac{2\pi nt}{T}\right) dt\right\}$$

$$= \frac{8}{T\pi n}\left[-\frac{T}{2\pi n}\cos\left(\frac{2\pi nt}{T}\right)\right]_{0}^{T/2}$$

$$= \frac{4}{\pi^2 n^2}(1 - \cos(n\pi))$$

that is, $a_n = 8/(\pi^2 n^2)$ when n is odd and $a_n = 0$ when n is even, and the Fourier series representation of a triangular wave may then be obtained using equation 4.10. Figure 4.8 shows the Fourier series representation of a triangular wave using up to ten terms. It can be seen that the series is rapidly convergent.

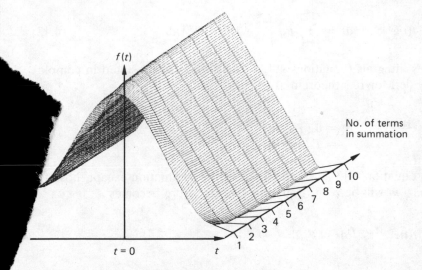

Figure 4.8   Fourier series representations of a triangular wave

4.4   THE COMPLEX FORM OF THE SERIES

Using de Moivre's theorem, one form of which can be expressed by the equations

$$\cos(x) = \frac{1}{2}\{e^{jx} + e^{-jx}\} \quad \text{and} \quad \sin(x) = \frac{1}{2j}\{e^{jx} - e^{-jx}\}$$

the real form of the Fourier series (equation 4.2) may be written:

$$f(t) = \frac{a_0}{2} + \sum_{n=1}^{\infty}\left\{\frac{1}{2}\left(a_n + \frac{b_n}{j}\right)e^{(j2\pi nt/T)} + \frac{1}{2}\left(a_n - \frac{b_n}{j}\right)e^{(-j2\pi nt/T)}\right\}$$

which by changing the limits of the summation [see reference [1] for details] is of the form

$$f(t) = \sum_{n=1}^{\infty} F_n e^{(j2\pi nt/T)} \tag{4.12}$$

where, for a real function $f(t)$ the coefficients F_n are given by $F_0 = a_0/2$ and $F_n = (a_n + jb_n)/2$.

Equation 4.12 defines the complex form of the Fourier series. A formula for the coefficients F_n may be derived directly. Both sides of 4.12 are multiplied by $e^{-j2\pi mt/T}$ and integrated from t_1 to t_2. This gives:

$$\int_{t_1}^{t_2} f(t)e^{-j2\pi mt/T}dt = \sum_{n=-\infty}^{+\infty} F_n \int_{t_1}^{t_2} e^{-j2\pi(m-n)t/T} dt \qquad (4.13)$$

Fourier's integrals (equations 4.4, 4.5 and 4.6) can be stated in complex form by de Moivre's theorem as

$$\int_{t_1}^{t_2} e^{j2\pi(n-m)t/T}dt = 0 \ (n \neq m)$$
$$= T \ (n = m)$$

Thus in equation 4.13 all the integrals in the summation except those t which $n = m$ will be zero. Equation 4.13 therefore becomes

$$\int_{t_1}^{t_2} f(t)e^{(-j2\pi mt/T)}dt = F_m T$$

If we replace m by n we obtain the expression

$$F_n = \frac{1}{T}\int_{t_1}^{t_2} f(t)e^{(-j2\pi nt/T)}dt \qquad (4.14)$$

which gives the values of all the coefficients.

4.5   THE FOURIER TRANSFORM

A restriction on the use of the Fourier series is that it can only be applied to periodic signals. Many real signals are not periodic. For example, consider the response of a tuning fork when struck, the sound of a gun firing, or the noise of the sea. The complex form of the Fourier series (equations 4.12 and 4.14) can be modified to deal with this situation as follows.

Suppose that the function $f(t)$ which we wish to represent by a Fourier series is a transient. As discussed earlier, an associated periodic function can be constructed which is identical to $f(t)$ within the limits t_1 to $t_2 = t_1 + T$, but which repeats itself indefinitely outside this interval. However, if we want to represent our function over the whole range of t, it is necessary to let $t_1 \to -\infty$ and $t_2 \to +\infty$.

The effect of increasing the length of the interval T is to pack the frequency axis more densely and to proportionately reduce the magnitude of the coefficients F_n. At the limit when $T \to \infty$ the function $f(t)$ will be correctly represented for all values of t, the separation of the terms in the spectrum will be zero, and the coefficients F_n will be infinitesimally small. However, if we omit the factor $1/T$ from equation 4.14 they will remain finite as $T \to \infty$, and give the form of the distribution of amplitude. It is this which constitutes the Fourier Transform of the function $f(t)$. It is a continuous curve rather than a set of discrete coefficients. In other words, the Fourier Transform of a non-periodic function $x(t)$ is a continuous spectrum. By a process similar to that described above a second transform equivalent to equation 4.12 and known as the reverse Fourie. can be defined, which enables a time domain function to be rec from frequency domain information.

The Fourier Transform is then defined by the following pair of in which correspond to equations 4.14 and 4.12 respectively:

$$S(f) = \int_{-\infty}^{\infty} x(t)e^{-j2\pi ft}dt \quad \text{(the forward transform)} \tag{4.1}$$

$$x(t) = \int_{-\infty}^{\infty} S(f)e^{+j2\pi ft}df \quad \text{(the reverse transform)} \tag{4.16}$$

The expression $e^{\pm j2\pi ft} = \cos(2\pi ft) \pm j\sin(2\pi ft)$ is known as the kernel of the Fourier Transform. $S(f)$ is known as the Fourier Transform of $x(t)$. $S(f)$ is in general complex, and contains amplitude and phase information for all the frequencies which make up $x(t)$ even though $x(t)$ is not periodic. Since $S(f)$ is in general complex, if we want to plot $S(f)$ as a function of frequency we either have to plot two graphs (one representing the real, one the imaginary parts of $S(f)$), or we have to resort to a three-dimensional representation such as the examples in figure 4.9.

$S(f)$ can be written in the form

$$S(f) = a(f) - jb(f) \tag{4.17}$$

where $a(f)$ is the real and $-b(f)$ the imaginary part of $S(f)$. If $x(t)$ is an even function, then by definition $x(t)\sin2\pi ft$ is odd. Hence

$$b(f) = \int_{-\infty}^{\infty} x(t)\sin2\pi ft\,dt = 0 \tag{4.18}$$

Thus, in the case of an even function, the complex transform $S(f)$ degenerates to a pure real function with no imaginary part. Similarly, it can be shown that if $x(t)$ is odd, $a(f) = 0$ and for an odd function $S(f)$ is purely

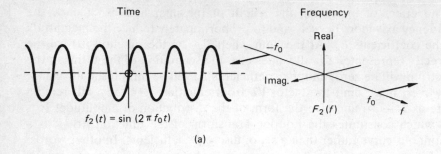

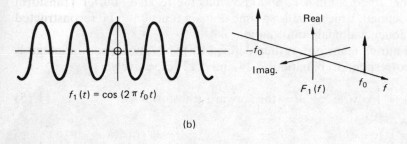

Figure 4.9 *Fourier Transforms of odd and even functions*

imaginary. Figures 4.9(a) and (b) show both cases using sine (odd) and cosine (even) functions.

These results are of great utility in evaluating the Fourier Transforms of various common functions.

Example

As an example, let us consider the Fourier Transform of a tone burst or cosine wave of short duration. This function arises in practice every time you switch on an oscillator and some time later switch it off again. In the time domain it has the form shown in figure 4.10.

The Fourier Transform of $x(t)$ in this case is

$$S(\omega) = \int_{-\infty}^{\infty} x(t)e^{-j\omega t}dt = \int_{-T/2}^{T/2} \cos Pt \, e^{-j\omega t}dt$$

where ω = angular frequency = $2\pi f$.

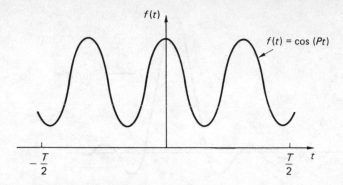

Figure 4.10

This is a standard integral which can be evaluated analytically, although the process is somewhat lengthy. However, recalling the properties of odd and even functions discussed above, we can take a short cut, since $x(t)$ in this case is obviously even. Therefore, its Fourier Transform is a pure real function, that is, it is equal to $a(\omega)$. Hence

$$S(\omega) = a(\omega) = \int_{-\infty}^{\infty} x(t)\cos\omega t\, dt$$

$$= \int_{-\infty}^{\infty} \cos Pt \cos\omega t\, dt$$

$$= \tfrac{1}{2} \int_{-T/2}^{T/2} \{\cos(\omega - P)t + \cos(\omega + P)t\}\, dt$$

$$= \frac{T}{2} \left\{ \frac{\sin\tfrac{1}{2}(\omega - P)T}{\tfrac{1}{2}(\omega - P)T} + \frac{\sin\tfrac{1}{2}(\omega + P)T}{\tfrac{1}{2}(\omega + P)T} \right\}$$

The number of cycles of the waveform in the pulse, that is, the number of cycles of angular frequency P occurring in time T, is given by $PT/2\pi$. If there are a large number of cycles in the pulse (PT is large) then the second term in the expression for $S(\omega)$ will be small for the positive range of ω. The curve for $S(\omega)$ then becomes the curve $\sin(x)/x$ centred at $\omega = P$ as shown in figure 4.11

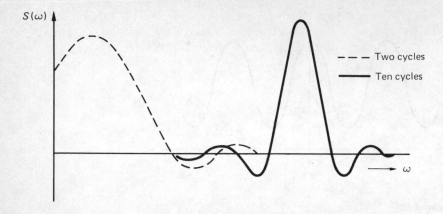

Figure 4.11

4.6 SUMMARY

In this chapter we have seen how periodic waveforms may be represented by either the real or the complex form of a Fourier series. Non-periodic waveforms may be represented within set limits by a Fourier series, but the series will also generate an associated periodic function outside the limits.

This problem may be avoided by the use of Fourier Transforms. The Fourier Transform integral is derived from the expression giving the coefficients F_n used in the complex Fourier series, but unlike the F_n coefficients (which are discrete) is a continuous analytic function. A Fourier Transform can be used to represent any deterministic function.

Most laboratory spectrum analysers use the Fourier Transform to convert a real time domain record into a complex frequency domain description. Usually either amplitude or phase is displayed as a function of frequency. The Fourier Transform is carried out digitally in a spectrum analyser. Problems may be encountered, since, as we have seen, the Fourier Transform is strictly a continuous analytic integral and its digital implementation necessarily involves sampling. The digital implementation of the Fourier Transform is the subject of the next chapter.

In noise measurement we are often interested in the frequency content of a sound rather than its overall level. The FFT allows the contributions made to a sound by different frequency components to be estimated. Frequency analysis is used for two reasons. First, it is essential in assessing the likely human response to a noise, since the human ear does not have the same sensitivity at all frequencies. Second, frequency analysis using the FFT permits very detailed examination of a spectrum. This is useful when, for example, analysing the noise emitted by a machine before designing noise reduction devices. The location of noise sources within devices which emit many distinct harmonics (such as gearboxes) and those which emit pure tones (such as gas turbines) is also very difficult without an FFT-based spectrum analyser.

REFERENCES/FURTHER READING

[1] R. D. Stuart, *An Introduction to Fourier Analysis,* Science Paper-backs, Methuen (1961).
[2] J. Bendat and A. Piersol, *Engineering Applications of Correlation and Spectral Analysis*, Wiley (1980).
[3] R. B. Randall, *Frequency Analysis*, 3rd edn, Bruel & Kjaer (1987).
[4] D. Brook and R. J. Wynne, *Signal Processing – principles and applications*, Edward Arnold (1988).

5. Digital methods of spectrum analysis

I have done the deed. Didst thou not
hear a noise?
Love's Labour's Lost, Shakespeare

5.1 INTRODUCTION

In the previous chapter we saw how acoustic (or any other) signals may be examined in either the time or the frequency domain by means of the Fourier Transform (the FT). The FT is a continuous analytic integral, which is calculated digitally in most modern spectrum analysers. The digital implementation of the Fourier Transform (defined by equations 4.15 and 4.16) necessarily involves sampling the data. Sampling can introduce undesirable effects such as aliasing, which may lead to erroneous results in the hands of the unwary user.

A second phenomenon, which limits the accuracy of a spectral estimation, is concerned with the relationship between the frequencies contained within a signal and the length of the time domain record of the signal. If, as is likely, the record contains a non-integral number of cycles of each frequency, an effect known as leakage will occur. In cases where two components are close in frequency, leakage often causes the smaller amplitude to be obscured by the larger. While this effect can never be totally avoided, its effects may be reduced by the technique of windowing.

The intention of this chapter is to discuss some of the pitfalls associated with practical spectrum analysis, together with means by which they may be avoided. An outline of the way in which the digital Fourier Transform is implemented is given first, since understanding the process is an essential prerequisite to using it intelligently.

5.2 THE DIGITAL FOURIER TRANSFORM AND THE FFT

We saw in chapter 4 how the Fourier Transform (equation 4.15) is used to obtain a frequency domain version of a continuous time domain signal. To

Fourier-Transform a time domain signal using a digital system, the continuous time signal must be represented by a set of discrete data points. The process is akin to writing down readings from a thermometer once every 10 minutes. The implicit assumption is that the temperature will not change dramatically in the 10 minute sampling interval, which is perfectly reasonable if, say, the temperature in a room is being recorded. In digital systems a sensor (such as a microphone) is used to provide an electrical analogue of the signal being investigated. Electronic circuits examine the sensor output at regular intervals, and convert the voltage or current to a digital value using binary code. The technical expression for this process is that the signal is sampled and digitised.

Equation 4.15 is however a continuous integral. If we have a signal represented by a set of discrete points, measured at intervals Δt, we cannot use equation 4.15 directly to obtain a spectrum of the signal.

One way of visualising a continuous function is to think of it as a set of data points separated by infinitesimally small intervals, that is, $\Delta t \rightarrow dt$. This is obviously impossible in a practical system. If we are to have finite-valued sampling intervals Δt, we must replace the continuous Fourier Transform integral by a summation as shown in equation 5.1:

$$S'(f) = \Delta t \sum_{n=-\infty}^{\infty} x(n\Delta t)e^{-j2\pi fn\Delta t} \tag{5.1}$$

$x(n\Delta t)$ is the value of the signal measured at the sampling intervals Δt. The summation shows that we can still calculate a valid Fourier Transform, even though we are dealing with a discrete number of samples measured at finite sampling intervals Δt, large compared with the infinitesimal dt.

However, the magnitude and phase information for all the frequencies contained in $S(f)$ is not accurate when a Fourier Transform is calculated in this manner. The function $S'(f)$ accurately describes the spectrum of $x(t)$ only up to a frequency F_{max}, where F_{max} depends on the sampling interval as discussed later (see Shannon's Sampling Theorem).

Examination of equation 5.1 shows that if $S'(f)$ is to be calculated accurately, an infinite number of samples of the input waveform will be required. Again, this is obviously impossible. Any practical spectrum analyser has to deal with a finite number of data points accumulated over a finite length of time. If the number of data points is N and the sampling interval is Δt, then the record length $T = N\Delta t$. Restricting the sampling time to finite values of T is equivalent to truncating the summation, and we cannot therefore calculate magnitude and phase values for an unlimited number of frequencies between 0 and F_{max}. The truncated version of the summation will not produce a continuous frequency spectrum, but is evaluated at discrete frequency points Δf Hertz apart. Each point represents the integral of the frequency domain representation over a frequency band of width Δf. A good way of visualising this is to consider the frequency domain representation produced by a digital Fourier Transform as a contiguous set of bandpass filters with bandwidth Δf, one filter for each point in the spectrum.

We can express the truncated summation produced by N data points as:

$$S''(m\Delta f) = \Delta t \sum_{n=0}^{N-1} x(n\Delta t)e^{-j2\pi m\Delta f n\Delta t} \tag{5.2}$$

$$(\text{for } m = 0, 1, 2, \ldots, \frac{N}{2}-1)$$

This expression is known as the Discrete Fourier Transform or DFT. It is a sampled Fourier series with $N/2$ real-valued time domain data points, and it can easily be implemented on a computer. An efficient algorithm for evaluating the DFT is the Fast Fourier Transform, or FFT. This algorithm, or one of its variants, is the basis of most digital spectrum analysers. A restriction on the use of the FFT is that it can only be used when N is a power of two, that is, when there are 256, 512, 1024 and so on data points. This condition does not apply to the DFT.

5.3 PRACTICAL USE OF THE FFT FOR SPECTRAL ANALYSIS

> *"Sir, are you so grossly ignorant of human nature,*
> *as not to know that a man may be very sincere in*
> *good principles, without having good practice?"*
> Dr Johnson to Boswell, *Tour to the Hebrides*

Let a continuous analogue signal from a sensor be sampled at regular intervals Δt, as shown in figure 5.1. If N samples are taken, the length of the data record must be $T = N\Delta t$.

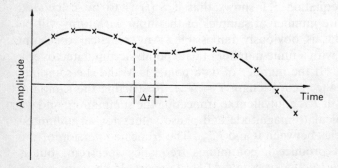

Figure 5.1

The data shown in figure 5.1 is in the time domain. It can be described in the frequency domain by means of a Fourier Transform, which will produce a set of frequency domain points defined by the following parameters:

Δf = frequency interval between points
$N/2$ = no. of frequency domain points
F_{max} = maximum frequency = $(N\Delta f/2)$

Note that in general, the following relationships between time and frequency domain data hold:

$$\Delta f = 1/T \tag{5.3}$$

$$\Delta t = T/N = 1/(N\Delta f) \tag{5.4}$$

It is important that the right choices are made when deciding how to digitise an analogue signal. This is particularly vital if no analogue record (such as a tape recording) is being made of the signal – a wrong decision at this stage usually means a wasted experiment!

For a particular application, table 5.1 can be used to help:

Table 5.1 Choice of sampling parameters.

Choose a convenient value for parameter shown	Chosen parameter fixes (because)	Make either of remaining 2 (not both) as convenient as poss. by choosing N
Δt	F_{max} ($F_{max} = 1/2\Delta t$)	T ($= N\Delta t$) Δf ($= 1/N\Delta t$)
T	Δf ($\Delta f = 1/T$)	Δt ($= T/N$) F_{max} ($= N\Delta f/2$)
F_{max}	Δt ($\Delta t = 1/2F_{max}$)	T ($= N\Delta t$) Δf ($= 1/N\Delta t$)
Δf	T ($T = 1/\Delta f$)	Δt ($= T/N$) F_{max} ($= N\Delta f/2$)

It is required that N (the number of points in the time domain) is always a power of two when using the FFT. The maximum value for N is normally fixed by the capabilities of the instrument you are using.

Example of the use of table 5.1

Suppose we want to examine the spectrum of noise produced by a gearbox. We wish to look at the frequency range up to 2 kHz with a resolution of 4 Hz. How many data points should we take?

Solution

We want $\Delta f = 4$ Hz. Therefore $T = 1/\Delta f = 0.25$ s.
$F_{max} = 2000 = N\Delta f/2$
If $\Delta f = 4$ Hz, N must be 1000 points.

In an FFT analyser where the number of data points is constrained to be a power of two, we should therefore use 1024 points ($= 2^{10}$), which would give us a slightly better resolution than we require.

5.4   ALIASING AND SHANNON'S SAMPLING THEOREM

The phenomenon of aliasing can cause errors when an instrumentation system is used which samples the signal at regular intervals. The effect is best demonstrated by example. Consider the temperature monitoring device shown in figure 5.2.

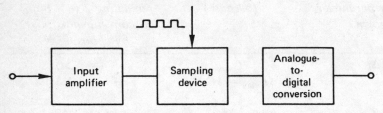

Figure 5.2

The system is set up to record the temperature in a room once a second. As the temperature of a room only changes slowly, we would expect each reading to be almost the same as the last, as shown by figure 5.3.

We are therefore sampling faster than is necessary in this case. Although no harm is done by sampling much more rapidly than the signal variations, the data storage capacity of the instrumentation used may be exhausted too quickly.

Alternatively, suppose we are measuring the pressure changes corresponding to the note 'middle C' (256 Hz). If we sample the signal from the microphone 256 times a second, the results will erroneously

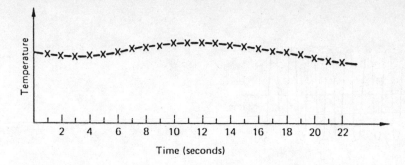

Figure 5.3

indicate that no sound is present, since we have sampled at the same point in each periodic pressure cycle.

This error is due to a phenomenon known as aliasing. A sinusoid and a sampling frequency are said to alias if the difference of their frequencies falls within the range of interest. An alias (or difference) frequency is always generated by the sampling process. However, if we sample at more than twice the highest frequency contained in a signal, the alias frequency will fall outside the frequency range of the input signal, and will not generate an error. In figure 5.4 the input frequency is greater than half the sampling frequency, so a low-frequency alias term is generated. Figure 5.5 shows aliasing in the time domain.

Aliasing is not always bad. Among other things it is used in a stroboscope to examine vibrating or rotating machines.

If we sample slower than twice the frequency of the input signal, a false low-frequency variation appears as shown by figures 5.4 and 5.5. When the sample rate is less than twice the highest frequency present in the input

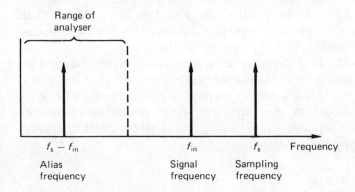

Figure 5.4

Temperature

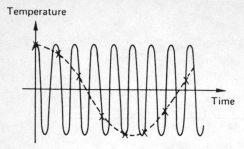

Figure 5.5

voltage, an alias frequency appears. This is expressed by Shannon's Sampling Theorem, one version of which states that to avoid these errors a sampled time signal must not contain components at frequencies above half the sampling rate (the so-called Nyquist frequency) [2]. If the frequency of the input voltage f_{in} is greater than half the sample rate f_s, the spurious signal that results will appear to have a frequency f_{alias} given by

$$f_{alias} = f_s - f_{in} \qquad\qquad (5.5)$$

The requirement for correct measurement, that the sample rate be faster than twice the highest frequency contained in the input signal, is known as the Nyquist Criterion. In the case of the room temperature measurement, we can be reasonably sure of the maximum rate at which the temperature might change. What we cannot be so sure of is whether the outside world will spoil things for us by introducing spurious signals into the system, for instance from mains wiring or from nearby radio transmitters. The only way to be quite certain that the input frequency range is limited is to add a low-pass filter before the sampler and the analogue-to-digital converter (ADC). Such a filter is often called the anti-alias filter. Many (but unfortunately not all!) spectrum analysers automatically apply anti-alias filtering to input signals before they are sampled.

A common mistake is to misinterpret Shannon's Sampling Theorem as meaning that you have to sample at more than twice the highest frequency *you are interested in*. This is wrong! You must sample more than twice as fast as the highest frequency *contained in the signal*. A good rule of thumb is always to use a sample rate five times higher than the highest frequency component in the filtered signal.

5.5 WINDOWING

> *"Accustom your children constantly to this; if a thing happens in one window and they, when relating it, say that it happened at another, do not let it pass but instantly check them; for you do not know where deviation from the truth will end."*
> Dr. Johnson, *Boswell's Life of Dr Johnson*

We saw in the previous chapter how truncation of a cosine tone burst affects its spectrum (see figure 4.11). To obtain a sharp peak in the magnitude spectrum, it is necessary to include as many cycles of the periodic function as possible in the record from which the Fourier Transform is to be calculated. However, the instrumentation system you are using limits this – no matter how expensive the spectrum analyser, it is never possible to take an infinite number of samples! In many cases the mechanical system will also limit the number of cycles available to be recorded, because of damping.

A further property of the FFT affects its use in frequency domain analysis. The FFT computes the frequency spectrum from a block of samples of the input, called a sampled time record. The Fourier Transform, to which the FFT is an approximation, is an integral extending over all time. Thus, the implicit assumption is made that the record contained in the sampled block is repeated throughout time.

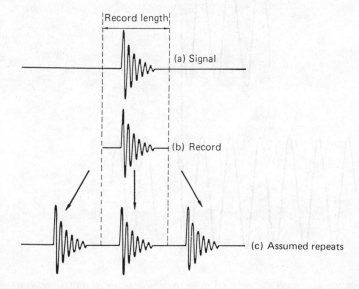

Figure 5.6

This does not cause a problem when we are dealing with a transient such as that shown in figure 5.6, since the 'join' between the repeats is smooth and will not distort the spectrum.

However, consider the case of a continuous signal such as a sine wave. If the time record is arranged (usually by a judicious choice of the sample rate) so that it contains an integral number of cycles, then the assumption of continuity exactly matches the actual input waveform, as shown in figure 5.7. In such a case, the input waveform is said to be periodic in the time record. The resulting magnitude spectrum is shown in figure 5.8.

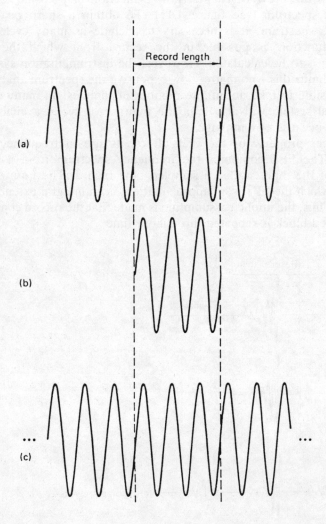

Figure 5.7 Assumed repeats match the original if the record is periodic within the data acquisition window

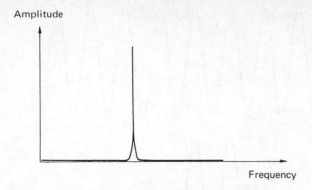

Figure 5.8   Spectrum obtained from data of figure 5.7

Figure 5.9 demonstrates the effect that occurs when an input is not periodic within the time record. The FFT is computed on the basis of the highly distorted waveform shown.

We saw in figure 4.6 that the (magnitude) spectrum of a continuous sine wave consists of a single line. The spectrum of the function shown in figure 5.9 will be very different. As a general rule, functions that are sharp or spiky in one domain appear spread-out in the other domain. Thus, we should expect the spectrum of our sine wave to be spread out throughout the frequency domain, and as shown in figure 5.10 this is exactly what happens. The power of the sine wave has been spread throughout the spectrum as predicted.

This smearing of energy throughout the frequency domain is known as leakage. Energy is said to leak out of one resolution line of the FFT into all the other lines.

It is important to realise that leakage is due to the fact that we have taken a finite time record. For a sine wave to have a line spectrum, it would have to exist for all time. If we had an infinite time record, the FFT would calculate the line spectrum exactly. However, since we are not willing to wait forever to measure a spectrum, we have to put up with a finite record. This will cause leakage, unless we can arrange for our signal to be periodic in the time domain.

It is obvious from figure 5.10 that the problem of leakage is severe enough to mask small signals, if they are close in frequency to the main sinusoid. A partial solution to this problem is provided by a technique known as windowing.

If we return to the problem of a sine wave that is not periodic in the time domain (figure 5.9), we see that the difficulty seems to be associated with the ends of the time record. The central part is a good sine wave. If the FFT can be made to ignore the ends and concentrate on the middle, we

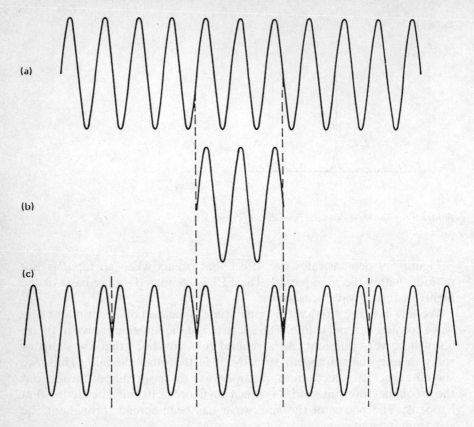

(a)

(b)

(c)

Figure 5.9   Input not periodic within time record, resulting in distorted 'assumed repeats'

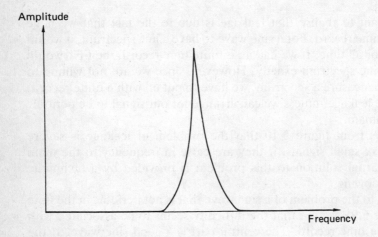

Amplitude

Frequency

Figure 5.10   Spectrum obtained from data of figure 5.9

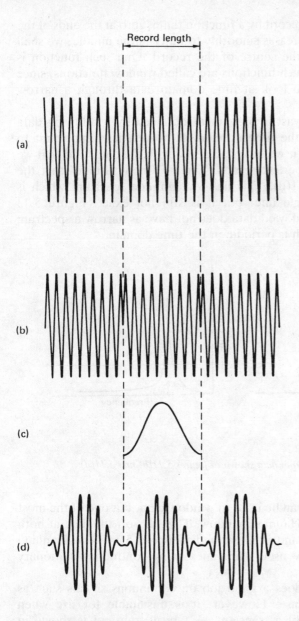

Record length

(a)

(b)

(c)

(d)

Figure 5.11 Windowing: (a) the signal; (b) assumed repeats; (c) the window function; (d) assumed repeats with window

would expect to get closer to the correct single line spectrum in the frequency domain.

If we multiply the time record by a function that is zero at the ends of the time record and which increases smoothly to unity in the middle, we shall concentrate the FFT on the centre of the record. One such function is shown in figure 5.11(c). Such functions are called window functions, since they force the analyser to look at time domain data through a narrow window.

Figure 5.12 shows the vast improvement obtained by windowing data which is not periodic in the time domain. However, it is important to realise that by windowing it we have tampered with the data, and that we cannot therefore expect perfect results. The FFT now assumes that the data looks like figure 5.11(d). This has a magnitude spectrum which is closer to the correct single line, but it is still not exact. Figure 5.13 demonstrates that the windowed data does not have as narrow a spectrum as unwindowed data which is periodic in the time domain.

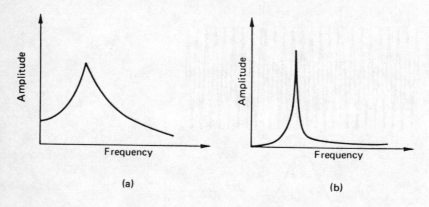

Figure 5.12   Spectra obtained from data shown in figures 5.11(b) and 5.11(d)

A number of functions can be used to window data, but one of the most common is known as the Hanning window. This tails-off the data at both ends in a cosine^2 shape, and is illustrated in figure 5.14. The Hanning window is also commonly used when measuring continuous stationary random noise.

The Hanning window does a good job on continuous signals such as sinusoids and random noise. However, it is unsuitable for use when calculating the spectrum of a transient. A typical transient is shown in figure 5.15(a). If the Hanning window shown in 5.15(b) is used on the data, we get the highly distorted signal shown in 5.15(c). If the transient represents, for example, the sound of a bell after it has been struck, all the vibration in higher modes occurs in the early part of the record and is likely

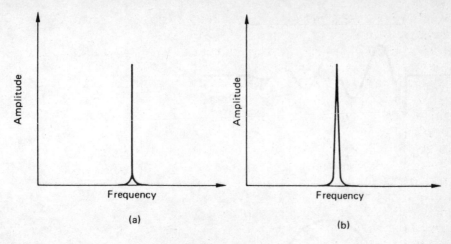

Figure 5.13 *Spectra from: (a) unwindowed data which is periodic within the acquisition period; (b) non-periodic data after windowing*

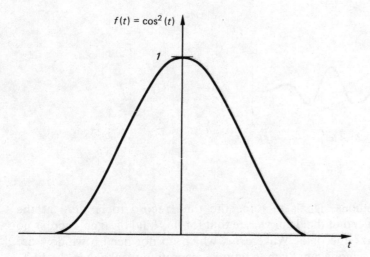

Figure 5.14 *The Hanning (cosine²) window*

to be lost after windowing. The spectrum of the transient, with and without a Hanning window, is shown in figure 5.16. The Hanning window has taken the transient, which naturally has energy spread widely through the frequency domain, and has distorted it until it resembles a sine wave. For transients, using a Hanning window is dangerous and should be avoided.

Many transients have the property that they are zero at the beginning and end of the time record. Recalling that we introduced windowing in the

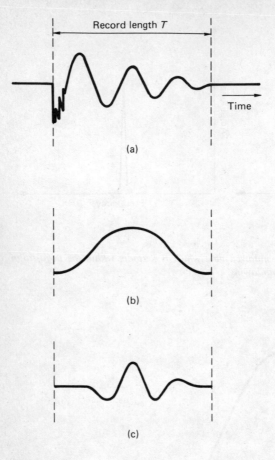

Figure 5.15

case of continuous signals to force the time record to be zero at the beginning and end of the data, we see that for this kind of signal there is no need to window the data. Waveforms which do not need a window are known as self-windowing. Such functions generate no leakage in the FFT. We can consider the data as being viewed through a rectangular or uniform window, created by our action of turning the recording instrument on and off.

One means of obtaining a finer resolution is to add zeros to the data record, but this is of course at the expense of a larger transform size and a longer processing time. For example, if a record of length N is extended to $2N$ with zeros, and a transform of size $2N$ performed, then the resolution will be $f_s/2N$ while the bandwidth will still be f_s/N. Another way of thinking about this is to imagine that we have applied a rectangular weighting function of length N to a data record of length $2N$.

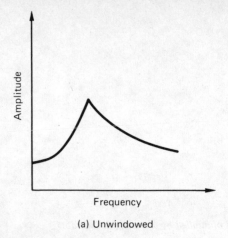

(a) Unwindowed

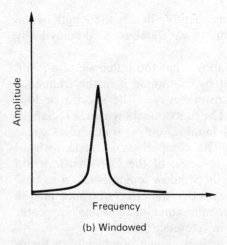

(b) Windowed

Figure 5.16

5.6 CHOICE OF WINDOW

> *The difficulty in life is the choice.*
> George Moore, *The Bending of the Bough*

The tone burst discussed earlier can be considered as an infinite sinusoid multiplied by a rectangular window, as shown in figure 5.17. The operation of multiplying two waveforms together in the time domain corresponds to convolution of the respective spectra in the frequency domain (see, for example, reference [1] for details). Changing the length of the rectangular

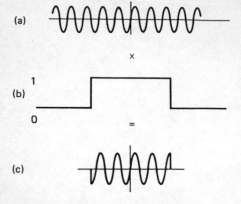

Figure 5.17 *Tone burst as an infinite sinusoid multiplied by a rectangular window*

window changes its spectrum, and consequently altering the length of the tone burst alters the resulting spectrum, as was demonstrated analytically in the last chapter.

It can be seen from the discussion above that the influence of a given window function can best be assessed by examining it in the frequency domain. Figure 5.18 shows a comparison between the spectra of four commonly used windowing functions. The functions shown are a rectangular window, Hanning, Hamming (Hanning on a small rectangular pedestal), and a Gaussian function. The comparison is made with all windows having the same length. In the case of the Gaussian function (which is theoretically infinitely long) the window length is taken to be 7 times the standard deviation, which means that the effect of truncating the window will not be seen unless the dynamic range of the signal is greater than 60 dB. Further details are given in reference [2].

Table 5.2 compares the window functions shown graphically in figure 5.18, and gives numerical values for the 3 dB bandwidth, the height of the largest sidelobe, and the rate of fall-off of the sidelobes. In each case the length of the window is T.

Table 5.2 Comparison of window functions.

Name	3 dB bandwidth	Size of largest sidelobe	Sidelobe fall-off rate
Rectangular	$0.9/T$	−13 dB	20 dB/decade
Hanning	$1.4/T$	−32 dB	60 dB/decade
Hamming	$1.3/T$	−42 dB	20 dB/decade
Gaussian	$1.8/T$	none	no sidelobes

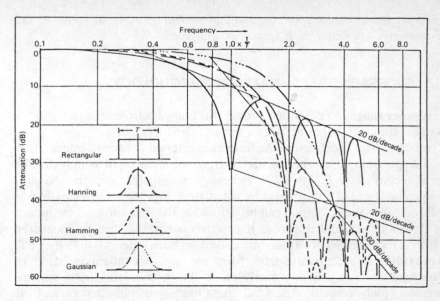

Figure 5.18   *Comparison of window functions in the frequency domain (courtesy of Bruel &*
Kjaer Ltd)

From table 5.2 and figure 5.18 it can be seen that the Hanning window
gives a much better performance than the rectangular function. A Hanning
window is generated by multiplying the time domain data by a raised
cosine function [1], which has a maximum value of 1 and is zero at both
ends. Because of this ease of generation, the Hanning window is available
on almost all analysers.

The Hamming window is produced by mounting a Hanning window on
top of a small rectangular pedestal. The first sidelobe of the Hanning
function coincides with and has opposite phase to the second sidelobe of
the rectangular function. This opposite phase can be made to cancel, by
appropriate choice of the height of the rectangular pedestal. The height of
the first sidelobe of the Hanning function is thus reduced in the Hamming
window. However, the subsequent sidelobes are dominated by the
rectangular function, and only fall off at 20 dB/decade, compared with 60
dB/decade for Hanning.

A well-known property of the Gaussian function is that its Fourier
Transform is another Gaussian function. The general form is $\exp(-x^2)$,
and when plotted on a decibel scale this appears as an inverted parabola
with no sidelobes. From this point of view it makes an ideal window
function. However, its bandwidth is greater than that of any of the other
functions shown. Where the length of a record is restricted due to
equipment or other limitations, probably a Hanning or Hamming window
is the best choice.

To complete this chapter we shall present a glossary of some of the terminology associated with spectrum analysis.

5.7   GLOSSARY OF FFT ANALYSER TERMINOLOGY

Frequency range. As described above, this is always from zero (DC) to the Nyquist frequency (that is, half of the sampling frequency).

Resolution. This is defined as the frequency interval between adjacent lines in the spectrum. It can be calculated from the number of samples in the time record: $R = f_s/N = 2f_n/N$ (where f_s = sample rate; the Nyquist frequency $f_n = f_s/2$). It can also be calculated from the record length T as follows. Since the resolution can be defined as the first non-zero frequency in the spectrum, and since the lowest frequency that can be measured within a record of length T must by definition have a period of T, $R = 1/T$.

Bandwidth. The term originates from the use of contiguous banks of bandpass filters as spectrum analysers in the days before digital systems became widely available. Ideal bandpass filters pass only that part of the total power whose frequency lies within a finite range (the bandwidth). The concept can be understood by considering an ideal filter, which transmits at full power all signals lying within its passband and completely attenuates all signals at other frequencies. The FFT can be thought of as a bank of ideal filters (one for each point in the spectrum), each having a passband or bandwidth determined by the sample rate and record size.

The bandwidth is determined by the resolution of the analyser, and by any time window applied to the data. For linearly weighted data the effective bandwidth is equal to the resolution, that is, $B_{eff} = R$. If any time window is applied to the data then the bandwidth of the result is equal to the bandwidth of the weighting function. In particular, the common Hanning window results in an effective noise bandwidth $B_{eff} = 1.5R$.

Dynamic range. This is determined by the number of bits with which the input data is represented. As a rough rule-of-thumb, 6 dB of dynamic range are obtained for every bit of the input data, and thus 72 dB is obtained from A/D conversion with 12 bits. The dynamic range is generally unaffected by the FFT calculation if 4 extra bits (that is, 16 bits for 12 bit data) are used for the arithmetic. Some increase in dynamic range can be achieved by averaging spectra in the frequency domain.

Zoom. The FFT algorithms discussed so far result in a so-called 'base band analysis', where the frequency range extends from zero up to the Nyquist frequency. It is sometimes useful to obtain greater resolution over a limited portion of the spectrum, and the zoom-FFT procedure permits this. The procedure involves complex manipulation of the data which space does not permit us to go into here. The interested reader will find details of the process in reference [2].

REFERENCES/FURTHER READING

[1] J. S. Bendat and A. G. Piersol, *Engineering Applications of Correlation and Spectral Analysis*, Wiley (1980).
[2] R. B. Randall, *Frequency Analysis*, 3rd edn, Bruel & Kjaer (1987).

6. Acoustic instrumentation

This particularly rapid, unintelligible patter isn't
generally heard, and if it is it doesn't matter.
Ruddigore, (Gilbert & Sullivan)

6.1 INTRODUCTION

An acoustic wave consists of a variation in pressure as time passes. To measure or analyse an acoustic waveform, it is usual to generate an electrical analogue of the pressure variation by means of a transducer. Instead of dealing directly with the pressure changes, we study a changing voltage or current which has the same behaviour with respect to time. If we want to generate a sound we also start with an electrical waveform, and again use a transducer to convert it into a pressure change.

A transducer is a device which converts energy from one form to another. Transducers are divided into two classes, input transducers, or sensors, and output transducers, of which the loudspeaker is an example. Loudspeakers convert a varying electrical current into mechanical motion which produces sound waves. Microphones receive sound waves, convert the pressure variation into mechanical motion, and use the motion to generate an electrical signal. In this chapter, which is concerned with sound measurement, we shall mainly be discussing microphones.

Once the electrical analogue of a sound wave has been obtained, the engineer has to consider the design of the rest of the instrumentation system. A typical microphone is a small device which can only extract tiny amounts of energy from a sound wave. Its output signal is at best a few millivolts, and the device will only provide a few microamperes of current. It is therefore necessary to use a high-impedance amplifier to increase the size of the signal to a usable level. Generally, the peak voltage after amplification is a few volts.

Amplitude is not the only thing the acoustic engineer must think about. The frequency content of the sound and its relation to the frequency response of the microphone must also be considered. The sensitivity S of a

116

microphone is usually characterised by the output voltage resulting from a unit pressure:

$$S = |V_{out}/p(0)| \tag{6.1}$$

S is measured by slowly sweeping the frequency of a sinusoidal sound over a wide range, while holding the amplitude constant. S always varies with frequency to a greater or lesser extent. Microphones where S is almost constant over a large frequency range are called flat-response devices. These are the most convenient to use. However, never assume a microphone has a flat response unless it is justified by your own measurements, or by detailed manufacturer's specifications.

Once a microphone and its associated amplifier have been used to give an electrical analogue of a sound, it may be necessary to carry out some form of analysis. This commonly takes the form of spectral analysis, in which the frequency content of the sound is calculated as discussed in chapters 4 and 5. Spectral analysis usually entails digitising the sound at regular intervals. It will be recalled from chapter 5 that if the digitisation process is carried out incautiously, aliasing errors will be introduced. For this reason, anti-aliasing filters are often used after the amplification stage.

As an alternative to digital spectrum analysis, analogue spectrum analysers are sometimes used. These operate by measuring the energy transmitted by narrow bandpass filters as shown later. Filter analysis has generally been superseded by FFT analysis. However there are cases, for example in constant percentage bandwidth analysis, where a filter based analyser has advantages over the FFT system.

In some cases an acoustic signal may be simultaneously measured and analysed. This process is called on-line analysis. However, it is usually preferable to make a permanent record of the signal, even if on-line analysis is being used. If a recording is available, many different forms of analysis may be performed on the same signal. In addition, a safeguard is provided against accidents in which data is lost because of instrumentation failure.

The basic instrument used for noise measurement is the sound level meter. This consists of a microphone, an amplifier, a set of weighting filters and a meter to indicate the output, packaged into one portable unit. A typical sound level meter is shown in figure 6.1. The instrument measures noise intensity without discriminating between the frequencies present in the sound (apart from the imposition of a weighting function as discussed in chapter 2). A microphone with a flat response is used. An electrical output is normally provided so that the amplified signal from the microphone may be recorded or fed to a spectrum analyser if simple intensity measurements are insufficient. The weighting filters incorporated in a sound level meter allow the user to select a flat frequency response or one which mimics that of the human ear.

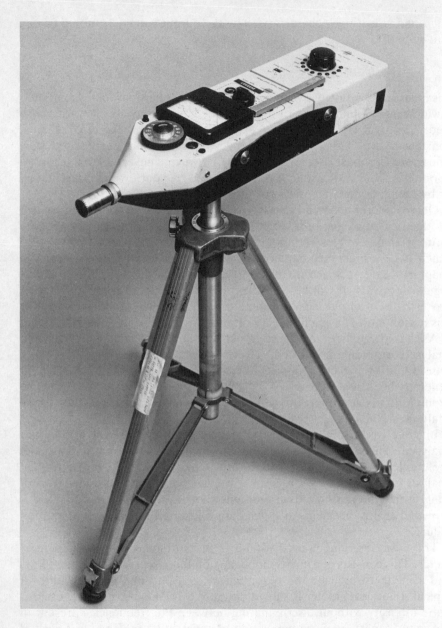

Figure 6.1   Sound level meter

6.2 MICROPHONES

As we saw in the introduction to this chapter, the microphone is the first (and probably the most important) element in the measurement chain. A good instrumentation microphone must be physically small, so as not to disturb the sound field it is trying to measure. It should also have a large acoustic impedance, so that it does not distort the sound field under investigation by extracting significant amounts of energy. Microphones should have a stable calibration and a wide frequency response. The directional properties should be known, and for a given frequency may be calculated from the microphone's dimensions as shown in chapter 3. If a microphone is to be used to measure low sound pressure levels (SPLs), it should have a low self-generated noise level. These requirements can best be met by capacitive or electret microphones.

There are three main types of microphone: capacitive, piezoelectric, and electromagnetic. Capacitor microphones are the most common and are generally used for laboratory/industrial noise measurement. Electro-magnetic microphones are still used for high-quality audio work such as broadcasting, although they are frequently replaced by capacitor micro-phones. Piezoelectric microphones do not offer such high performance as the other two, but are often used in domestic audio equipment because of their low cost. Because of their importance, capacitive microphones are discussed first and in most detail.

When selecting a microphone, the characteristics of the sound field to be measured should be borne in mind. The response of a microphone at high frequencies is influenced by the reflections and diffraction caused by its presence in the sound field. The response depends to some extent on the direction of the incident sound. Two main kinds of acoustic conditions exist: diffuse and free-field. If the field is diffuse, the sound is equally likely to arrive at the microphone from any direction. In a free field (the condition under which most open-air and many indoor noise measure-ments are made), the sound arrives predominantly from one direction. Care must be taken to match the microphone characteristics to those of the field being investigated, or the high-frequency response will suffer. Micro-phone characteristics are expressed in three ways: free-field, pressure, or random-incidence response.

A free-field microphone has a pressure/frequency response tailored so that it gives a flat response to the sound waves that would exist if the microphone were not present in the sound field. Free-field microphones are designed to be used with normally incident sound waves.

A pressure microphone has a uniform response to the sound field as it exists, including all disturbances caused by the presence of the micro-phone. Pressure microphones may be used for audiometer calibration or for free-field work if held at grazing incidence (that is, at 90° to the direction of incidence).

A random incidence microphone has a uniform response to sound arriving from any direction, and is therefore used when investigating diffuse fields.

Capacitive microphones

The capacitor microphone (also called a condenser or an electrostatic microphone) is essentially a parallel-plate capacitor. One of the plates is a thin membrane exposed to the medium in which the sound is to be measured, so that pressure fluctuations alter the capacitor plate spacing. The resulting capacitance changes cause fluctuations in the voltage across the capacitor. The output signal consists of a varying voltage V_{out} as shown in figure 6.2.

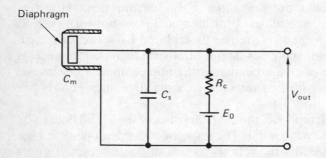

Figure 6.2   Equivalent circuit of capacitor microphone

The charge on the capacitor may be generated by an externally applied voltage, or by the properties of the material used to manufacture the capacitor. In the latter case, the sensor is often called an electret microphone. An electret is a permanently charged insulating material made by allowing a special molten plastic to solidify under the influence of a strong electric field.

The components of a typical capacitor microphone are shown in figure 6.3. A diaphragm made from metal foil is fixed close to an insulated rigid metal back plate. These two form a parallel-plate capacitor. A polarisation voltage E_0 is applied across the plates as shown in figure 6.2. The polarisation voltage source has a high impedance R_c, so that the time constant $R_c(C_m + C_s)$ is long compared with the lowest sound frequency to be measured. A further reason for making the time constant long is that this ensures the charge stored on the capacitor is approximately constant. If the charge is constant and the capacitance varies, a varying voltage will

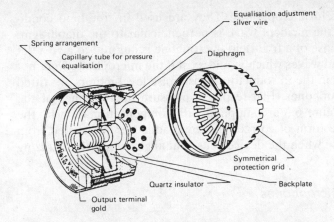

Figure 6.3   Sectional view of capacitor microphone (courtesy of Bruel & Kjaer Ltd)

appear across the plates. The capacitance C_s is due to unavoidable stray capacitance within the sensor, rather than a deliberate design feature.

If the sound pressure acting on the diaphragm produces a capacitance change ΔCm, then the output of the microphone will be a voltage V_{out} where

$$V_{out} = \frac{\Delta C_m E_0}{C_m + C_s} \qquad (6.2)$$

since $C_m \gg \Delta C_m$. Note that the microphone sensitivity S is proportional to the polarisation voltage E_0 but inversely proportional to the total capacitance $C_m + C_s$. Capacitor microphones can be manufactured with sensitivities as high as 100 mV/Nm^{-2} [1].

If we want the microphone to have a flat response, then the capacitance change ΔC_m and hence the deflection of the diaphragm for a given sound pressure must be independent of frequency. In other words, the diaphragm must be 'stiffness controlled' (see figure 1.9), and it will have a natural frequency well above that of the highest frequency sound to be measured.

The voltage output from a capacitor microphone is proportional to the applied sound pressure. However, when the frequency of the sound increases to the point where the wavelength is of the order of the diaphragm diameter, the diaphragm acts as a high-impedance obstacle in the sound field, from which the wave will be reflected. When this happens, the pressure sensed by the diaphragm is incorrectly high. This is a manifestation of the 'pressure doubling' effect discussed in chapter 7. In such a situation, a microphone with a flat pressure response would give an incorrect reading. For this reason, special capacitive microphones are made, called free-field microphones. They are used in free-field conditions, that is, when the incident wave is perpendicular to the diaphragm. The frequency response of a free-field microphone is tailored to give a flat response to the sound waves which would exist if the microphone were not present as discussed in the introducton. Most noise level meters are fitted with free-field microphones. Flat-response pressure microphones of the type illustrated in figure 6.3 are supplied with compensation data by the manufacturer, in the form of corrections to be added or subtracted from the pressure response when the device is used at angles other than grazing (that is, 90°).

Electromagnetic microphones

If a conductor carrying a current is placed in a magnetic field, it will experience a force perpendicular to both the current and the field. Conversely, a current will be generated in a conductor which moves perpendicularly through a magnetic field. These effects are used in the construction of both loudspeakers and microphones. In a magnetic microphone, pressure changes cause motion of a diaphragm carrying a small coil. The coil is arranged so that it lies within the field created by a permanent magnet. Movements of the diaphragm therefore cause a small voltage to appear across the ends of the coil. This type of microphone can be made extremely rugged, and it is therefore popular for sound recording and live performance use. However, magnetic microphones are usually heavy, because of the need for a permanent magnet, and have a relatively low sensitivity.

The system may be operated in reverse if a current is passed through the coil. A force is produced which actuates the diaphragm. This is the basis of the common dynamic loudspeaker. The diaphragm is replaced by a cone in most loudspeakers to improve the stiffness/mass ratio of the device.

Piezoelectric microphones

The piezoelectric effect occurs when a force is applied to a crystalline material such as quartz, producing an electrostatic charge, as shown in figure 6.4, which can be measured by a charge amplifier. Piezoelectric transducers can also be manufactured from ceramic materials which have greater sensitivity than quartz-based devices. In a piezoelectric microphone, acoustic pressure changes cause motion in a thin, stiff diaphragm, which applies stress to a piezoelectric material.

Crystalline piezoelectric microphones are manufactured from a single crystal of quartz or ammonium dihydrogen phosphate (ADP). Ceramic piezoelectric materials are polycrystalline, and include lead zirconate titanate (PZT) and barium titanate. Ceramic piezoelectric materials have the advantage that they can be moulded into any desired shape, and for this reason as well as their inherently higher sensitivity are more common than single crystal devices. However, their sensitivity can vary with time.

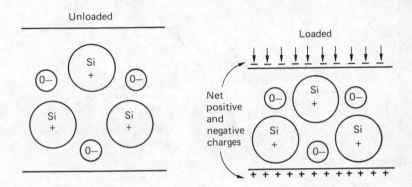

Figure 6.4   The longitudinal piezoelectric effect

6.3   MEASUREMENTS IN AN AIRSTREAM

Many outdoor measurements have to be taken in less than ideal conditions, for example in the presence of wind. If a microphone or any other obstacle is placed in a moving airstream, it will produce turbulence. Turbulence manifests itself as rapidly changing pressure around the microphone, which generates a spurious signal superimposed on the noise being measured. Figure 6.5 shows the magnitude of wind noise effects for microphones facing and at right angles to the wind, with and without a windshield, as a function of wind velocity.

The reduction in wind noise that can be achieved by the use of a suitable windshield is clearly shown on figure 6.5. The effect is most marked at velocities below 40 km/h. It is very difficult to make accurate noise measurements above this speed. Wind noise occurs mostly at the low-frequency end of the spectrum, so A-weighted measurements (see chapter 2) are less likely than others to be affected by wind noise since the low end of the spectrum is heavily attenuated by the A-weighting.

Windshields usually take the form of a sphere or oblong of foam material, sometimes supported on a wire frame. For special conditions where the airflow is of high speed and in a well-defined direction, smooth nose-cones are used. These give a low resistance to the airflow and reduce the turbulence. Figure 6.6 shows a selection of typical windshields and nosecones.

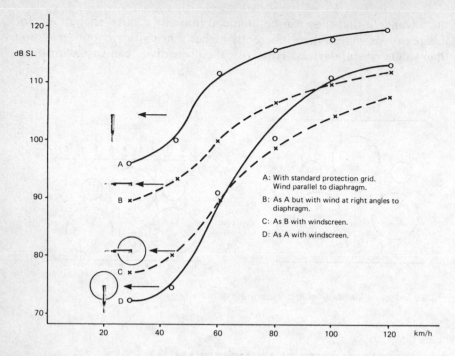

A: With standard protection grid.
   Wind parallel to diaphragm.

B: As A but with wind at right angles to
   diaphragm.

C: As B with windscreen.

D: As A with windscreen.

Figure 6.5 *Wind-induced noise as a function of wind speed. Measured with 1" B&K microphone (courtesy of Bruel & Kjaer Ltd)*

Figure 6.6 *Microphone windshields and nosecones*

6.4 HYDROPHONES

Sensors used to detect sound in liquids are called hydrophones since the transmission medium is usually water. The differences between a hydrophone and a microphone arise principally from the fact that water is roughly 1000 times denser than air. In a microphone for use in air, a thin diaphragm is needed to detect motion, since the amount of acoustic power incident on a microphone's surface is small. As water is more dense than air, a hydrophone is exposed to much larger forces. For this reason, hydrophones do not usually use a diaphragm to detect pressure changes, but use either the piezoelectric or the magnetostrictive effects for transduction. The working face of a piezoelectric or magnetostrictive hydrophone is (apart from a waterproof coating) exposed directly to the liquid medium.

Unlike acoustic measurements in air, where often the results are required in a form related to the frequency response of the human ear, some underwater work (such as sonar detection) is conducted at a single frequency. It is often unnecessary therefore for a hydrophone to have a flat response, and resonant sensors designed to give a strong response close to a fixed frequency are common.

6.5 MICROPHONE AND HYDROPHONE CALIBRATION

Microphones and hydrophones are individually calibrated after manufacture, and are supplied with a calibration chart such as the example shown in figure 6.7. This specifies the microphone's sensitivity over its entire frequency range. The principles of calibration are common to both microphones and hydrophones.

The sensitivity of a microphone should be checked at regular intervals to ensure that it remains within its specifications. If a microphone is found to have changed its sensitivity markedly, it is usually an indication that it has suffered some form of damage.

Reference and reciprocal calibration methods

The simplest and most direct way to calibrate a microphone is to compare its sensitivity with that of a reference device. A sound field of known characteristics is produced and the responses of the standard and the unknown device are compared. Care has to be taken to ensure that both microphones experience the same sound field conditions (such as incidence). The drawback in this method of calibration is that it requires a standard reference microphone.

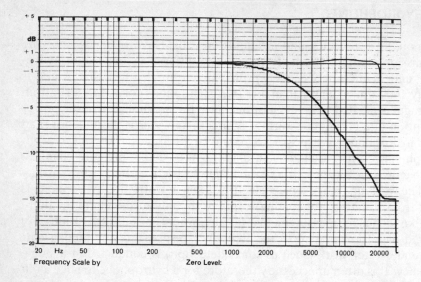

Figure 6.7   *Typical microphone calibration chart (courtesy of Bruel & Kjaer Ltd)*

Most microphones and loudspeakers are to some extent reversible, in that a loudspeaker can operate as a microphone and vice versa. This fact is used in an absolute calibration method which requires no reference standard. The technique makes use of the reciprocity theorem [2], which may be stated in the following form:

If a generalised force with magnitude *F* is applied to a linear system at a point A and the response *R* measured at point B, the ratio *R/F* (called the transfer impedance) will be unchanged if A and B are exchanged.

The sensitivity of a microphone was defined in equation 6.1 as

$$S = \left| \frac{V_{out}}{p(0)} \right|$$

Under most conditions this also equals the ratio of the volume velocity (that is, the volume of fluid displaced per unit time, units m^3/s) to the driving current if the transducer is reversed and driven as a loudspeaker. Hence

$$S = \left| \frac{V}{p} \right| = \left| \frac{Q}{i} \right| = R \tag{6.3}$$

where p is the sound pressure level (SPL) applied to the microphone, V the resulting open-circuit voltage, and Q the volume velocity of the transducer when it is driven as a loudspeaker by current i. Notice that the quantities V, p, i and Q are complex and that the modulus of each is used throughout this analysis.

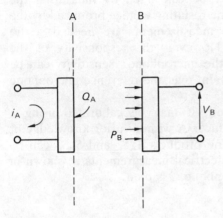

Figure 6.8   Reciprocity notation

If a reversible transducer at position A (see figure 6.8) is radiating sound with a volume velocity Q_A, and the resulting pressure is measured by a microphone at point B is p_B, then

$$|Q_A| = R_A|i_A| \tag{6.4}$$

and

$$|p_B| = \left|\frac{V_B}{S_B}\right| \tag{6.5}$$

where R_A and S_B are the response and sensitivity of the loudspeaker and microphone respectively. If equation 6.5 is divided by 6.4 we get:

$$\left|\frac{p_B}{Q_A}\right| = |Z_{AB}| = |Z_{BA}| = \left|\frac{V_B}{R_A S_B i_A}\right| \tag{6.6}$$

Note the reciprocity principle implies that acoustic transfer impedance is the same in either direction, that is $|Z_{AB}| = |Z_{BA}|$.

Equation 6.6 may be rewritten as

$$R_A S_B = \frac{|V_B|}{|Z_{AB}||i_A|} \tag{6.7}$$

Hence if the transducer impedance Z_{AB} is known, the product of the transducer sensitivities/responses may be calculated by measuring the current driving the loudspeaker and the resulting voltage produced by the microphone. Note that no pressure measurements are needed so no reference microphone is required. However, the responsivity of the loudspeaker must be known before the microphone's sensitivity can be calculated, so no significant improvement over the reference microphone method is obtained thus far.

If a third microphone C is added, it may be calibrated using a microphone B and a reversible transducer A without prior knowledge of any of the transducer sensitivities. The products $R_A S_C$ and $S_B S_C$ can be calculated from the impedance and electrical measurements as shown in equation 6.7. The sensitivity of microphone C is then

$$S_C = \sqrt{\left(\frac{R_A S_C \times S_B S_C}{R_A S_B} \right)} \tag{6.8}$$

and S_C can be calculated without knowing R_A or S_B. Further details are given in reference [2].

If the calibration is carried out in a cavity which is much smaller than the wavelength of the sound used, then

$$|Z_{AB}| = \frac{\gamma P_0}{\omega U}$$

where P_0 is the atmospheric pressure, γ is the ratio of specific heats and U is the volume of the cavity.

Three-transducer reciprocal calibration can also be carried out in free-field conditions (that is, when the distance between the transducers is much greater than the wavelength). In a free field the impedance is

$$|Z_{AB}| = \frac{\rho c}{2\pi r}$$

where ρ is density, c the wave speed and r the separation of the transducers, which are placed at the corners of an equilateral triangle.

The advantage of reciprocal calibration using three transducers is that no measurable or calculable sound pressures have to be produced. All the basic measurements (apart from distance if free-field calibration is used) are electrical.

Pistonphone calibration

For the purpose of obtaining calibrated measurements in the field, complex methods such as the use of a reference microphone or reciprocity calibration are not justified. Simpler techniques in which a known sound pressure level (SPL) at a fixed frequency is applied directly to a microphone are usually sufficient to ensure that a properly calibrated measurement is made. Portable battery-powered devices (known as pistonphones) are used in which a small electric motor or a piezoelectric driving element are used to vibrate a piston or diaphragm at one end of a cavity. The microphone is introduced through a seal at the other end of the cavity. The sound pressure produced is proportional to the ratio of the swept volume divided by the total volume of the cavity. Changes in atmospheric pressure produce small (less than 1 dB) changes in SPL. These can be compensated for by barometric measurements. The cavity is small compared with the wavelength of the sound used, so that the pressure field is uniform throughout the cavity (that is, the microphone is in the near field as discussed in chapter 3).

6.6 SOUND INTENSITY MEASUREMENT

> *The excellence of every act is its intensity, capable*
> *of making all disagreeables evaporate from their*
> *being in close proximity with . . .truth.*
> Keats, *Letters*

A technique developed in recent years which is rapidly gaining importance is sound intensity measurement. Unlike sound pressure measurements using a microphone, which result in (scalar) pressure values, intensity measurements are directional since intensity is a vector quantity. Thus measurements of intensity are more useful than those of pressure for locating the source of a noise and for noise control. The development of practical sound intensity measuring systems [3] in the last decade has led to the growing significance of this type of acoustic measurement.

Sound sources radiate power. The **effect** of the radiated power is a sound pressure level or SPL. An electric heater is also a power source, which in this case radiates heat. The **effect** of the heater is to raise the temperature of its surroundings. The analogy between the effect of the heater, temperature, and the effect of a sound power source, sound pressure, is a good one.

The temperature in a room containing the heater depends on the size of the room, its insulation properties, and whether or not other heat sources are present. The power radiated by the heater however is constant. Similarly, the relationship between sound power and SPL depends on the surroundings. The same acoustic power can give rise to entirely different pressure levels. This is just as well, or noise control measures would be doomed to failure.

Annoyance or even hearing damage may be caused by excessive sound pressure levels. In trying to assess the human response to noise, therefore, it is sensible to measure pressure, and as we have seen this is easily achieved by a diaphragm microphone.

A measured SPL is dependent on the acoustic environment and on the source distance, and can give no information about the power of the source. This is obvious – a loud machine in an absorbent environment may generate the same SPL as a quiet machine surrounded by reflective walls. Sound power is however independent of the measurement environment, and gives a unique description of the 'noisiness' of a sound source.

Sound power is the rate at which energy is radiated, that is, the work done (or energy) per unit time in Watts. Sound intensity is simply the power flowing through a unit area, which has the units W/m^2.

Sound intensity is a vector quantity since an energy flow has direction as well as magnitude. By contrast, pressure is a scalar quantity which only has magnitude. By convention, intensity is measured in the perpendicular direction to a specified unit area through which the energy is flowing.

Sound intensity is measured in the form of a time-averaged rate of energy flow per unit area. In some cases (such as near-field measurements) energy may be travelling back and forth with no net flow occurring. This will produce a time-averaged sound intensity of zero.

Intensity follows the inverse square law for far-field propagation from a spherical source. This may be seen by considering the propagation of sound from a symmetric spherical source as shown in figure 6.9. At distance r the source is enclosed by a sphere of surface area $4\pi r^2$. At distance $2r$ the sphere has area $16\pi r^2$. Since the power radiated must be the same in both cases, the intensity (power per unit area) must decrease by $1/r^2$. The intensity measured at $2r$ from the source will be four times smaller than that measured at distance r. (This also follows from the material presented in chapter 3.)

As we have seen, SPL measurements are essential for assessing the effects of noise on humans. However, if the purpose of the measurement is noise source location or noise reduction, intensity measurements may be preferable. Before the development of intensity probes, it was only possible to measure sound pressure using microphones. To assess the power of a sound source, it was necessary to measure its SPL under carefully controlled conditions, such as in an anechoic or reverberant

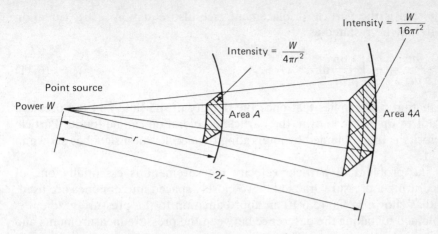

Figure 6.9   The intensity in a free field follows the inverse square law

chamber. This is unnecessary for intensity measurements which can be undertaken in any steady sound field. The measurements may be made *in situ* on an individual machine even when others are radiating noise, since diffuse background noise will make no contribution to the time-averaged intensity measurement.

Intensity probes

Sound intensity is the time-averaged product of pressure and particle velocity [3]. The pressure measurement is carried out by a microphone as in *SPL* estimation. However, the advances which led to the widespread adoption of intensity measurement were the development of a simple and reliable method for sensing particle velocity together with real-time signal processing techniques.

Euler's equation (6.9) is essentially Newton's second law applied to a fluid. Newton's law relates the acceleration of a mass to the applied force. Euler's equation relates the acceleration of a fluid with density to the pressure gradient experienced by the fluid. It can be stated as

$$\text{acceleration } a = -\frac{1}{\rho}\,\mathbf{grad}(p) \tag{6.9}$$

For one-dimensional flow this becomes

$$\frac{\partial^2 u}{\partial t^2} = -\frac{1}{\rho}\frac{\partial p}{\partial r} \tag{6.10}$$

where u is the particle displacement (see also equation 3.10). Equation 6.10 may be restated as

$$\frac{\partial u}{\partial t} = -\int \frac{1}{\rho}\frac{\partial p}{\partial r}\,\mathrm{d}t \qquad (6.11)$$

Equation 6.10 implies that once the density of the fluid and the pressure gradient $\partial p/\partial r$ are known, the particle acceleration can be found. Particle velocity is then obtained by integrating the acceleration signal as shown in 6.11.

The problem of particle velocity measurement is essentially one of measuring a pressure gradient. Two closely spaced microphones are used, and as shown in figure 6.10 an approximation to the pressure gradient is obtained by taking the difference between the pressure measurements and dividing by the spacing. The pressure gradient signal is integrated to give the particle velocity. The estimate of particle velocity is for the position midway between the two microphones. A pressure estimate is also obtained at the mid-point from the average pressures recorded by the two microphones. The pressure and particle velocity signals are multiplied together, and time averaging gives the intensity.

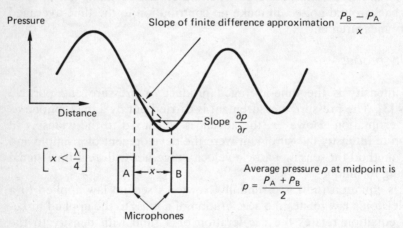

Figure 6.10   *Finite difference approximation used to obtain particle velocity from Euler's equation*

The above time domain approach to intensity measurement is perfectly feasible and has been used, but in practice the intensity is usually obtained by means of an FFT analyser. In the frequency domain the intensity is found to be proportional to the imaginary part of the cross-spectrum of two pressure signals. (A cross spectrum is the product of the Fourier Transform of a signal $x(t)$ and the complex conjugate of the Fourier Transform of a signal $y(t)$. See reference [7] for further details.) See reference [3] for more details of the different types of intensity probe.

The directional properties of a sound intensity probe

A sound intensity probe consists of two closely spaced microphones as described above. Usually the microphones are mounted face to face and separated by a solid spacer as shown in figure 6.11. The directional sensitivity of a sound intensity measurement system has a cosine characteristic as shown in figure 6.12 (assuming the spacing is much less than the wavelength). This is partly due to the probe geometry and partly to the signal processing methods used.

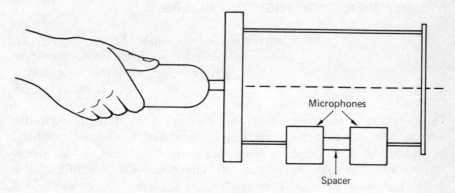

Figure 6.11   Sound intensity probe

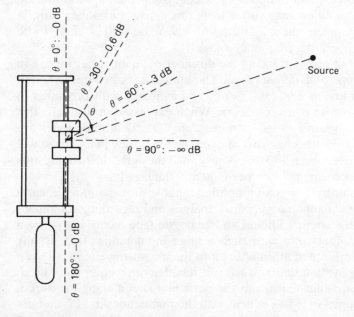

Figure 6.12   Directional characteristics of intensity probe

As we have seen, sound intensity is a vector quantity. However, with a two-microphone probe we do not measure the intensity vector but its component along the probe axis. For sound incident at 90° to the probe, there is no axial component (since cos 90° = 0). There will be no difference in pressure at the two microphones, and hence no axial particle velocity and no intensity in the direction of the probe. For sound incident at an arbitrary angle θ to the probe axis, the intensity component along the axis will be reduced by cos θ.

6.7   AMPLIFIERS AND WEIGHTING FILTERS

"But, in case signals can neither be seen or perfectly understood, no captain can do very wrong if he places himself alongside the enemy."
Horatio Nelson

The electrical output from a microphone is very small, as discussed in the introduction. Before it can be recorded or analysed, it must be amplified. In the ideal world an amplifier has a uniform gain and increases the signal amplitude by the same amount at all frequencies. Real amplifiers have gains which are to some extent dependent on frequency. In a well-designed system, however, the amplifier gain will be reasonably constant over the frequency range of the microphone used.

An ideal amplifier will not impose a frequency-dependent phase shift on the signal. Real amplifiers may suffer from this defect, but once again, in well-designed equipment the magnitude of any phase shift is likely to be small within the operating frequency range.

The most difficult problem facing the designer of an amplifier for use in acoustic work is that of dynamic range. The human ear can detect a sound pressure level as low as 2×10^{-5} N/m^2, and can cope with pressures to around 30 N/m^2 before damage occurs. When expressed in decibels, this corresponds to a dynamic range of more than 120 dB. Microphones are designed to work over the same sort of dynamic range, and in fact most will respond linearly over about 140 dB. Above this, they give distorted results and, like the human ear, may be permanently damaged.

It is very difficult to design an amplifier capable of responding linearly over such a large dynamic range. Most analysis and recording equipment (such as spectrum analysers, digital and analogue tape recorders, or even simple meters) cannot cope with such a large input range. It is usually necessary to put calibrated attenuators into the measuring circuit, to keep the dynamic range within limits which suit the electronic equipment used. The attenuators are adjusted to suit the particular sound being measured, so that the electronics only has to deal with the fluctuations in SPL and not with the constant background level. For example, if a noise varying between 70 and 80 dB were being investigated, a −70 dB attenuator might be used. The amplifier would then only need a dynamic range of 10 dB. The reading from a typical noise meter is obtained as the sum of the attenuation (usually given by the position of a rotary switch) and the reading on a meter mounted on the instrument.

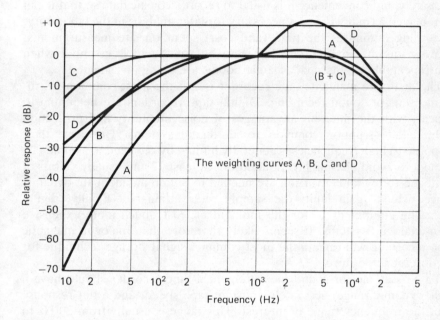

Figure 6.13   A-, B-, C- and D-weightings

Sound level meters usually combine the function of amplification with that of applying a weighting to the signal. There are three internationally agreed standard weighting functions (known as A, B, and C), which allow a sound level meter to mimic the sensitivity of the human ear, as discussed in chapter 2. Essentially a band-pass filter [4] is used which attenuates low and high frequencies by differing amounts. The A-, B- and C-weightings are shown in figure 6.13. A D-weighting is also sometimes used to estimate the perceived noisiness of aircraft in flight. Loudness readings obtained with weighting circuits are often quoted in phons (see chapter 2). Strictly, the A-weighting should be used for sound levels below 55 dB, the B-weighting between 55 and 85 dB, and the C-weighting above 85 dB. However, it is common to find the A-weighting used for all noise levels. Above 85 dB, A-weighted measurements should not strictly be quoted in phons. It has been found that in practice there is a reliable relationship between readings obtained using the A-weighting at all noise levels and the perceived loudness of a noise. Such measurements are usually quoted in dBA units.

6.8   SIGNAL STORAGE EQUIPMENT

For safety and convenience it is usual to record acoustic data so that it can be replayed as many times as necessary for later analysis in the laboratory. Recording minimises the time (and cost) spent on site measurements, allows different analysis techniques to be tried, and is essential when non-repeatable events such as sonic booms are being investigated.

The analogue magnetic tape recorder is still the most common instrument used for sound recording. Digital tape recorders are beginning to appear, with the inherent advantages of noise immunity and protection from data corruption common to all digital systems. However, their response to high frequencies is generally limited by the need to sample fast enough to avoid aliasing [4]. In addition, at least 12 and usually 16 bits of analogue-to-digital conversion are needed to obtain an adequate dynamic range, which again limits the sample rates attainable. Finally, digital recording equipment is, for the moment at least, much more expensive than analogue systems. It seems likely therefore that analogue magnetic tape recording will remain the most common signal storage technique for the foreseeable future.

To be suitable for sound measurement a tape recorder should have a wide dynamic range, accurately constant tape speed, and a flat response over the frequency range of interest. This range is usually from 20 Hz to around 20 kHz for audio work, but may extend to much lower frequencies for measurement of sonic booms and pressure pulses resulting from explosions.

Two techniques are used in tape recording. In direct recording a changing current (the signal) is converted to a varying magnetic field by a transducer. The field variation causes permanent changes in the magnetisation of iron particles carried past the recording head on a moving tape. This method may perform poorly at low frequencies, since the magnetic field changes will be slow and may not be properly recorded. Direct recording is generally only used for frequencies above 100 Hz.

Frequency modulated (FM) recording is most commonly used for sound measurement. In this approach, the signal is used to frequency-modulate a carrier signal. The modulated signal is recorded on tape, thus allowing frequencies down to DC to be recorded. The carrier frequency used depends on the tape speed and is quoted in Hz per inch per second of tape speed. 3600 Hz/ips is typical. The instrument shown in figure 6.10 has a frequency response from DC to 20 kHz at a tape speed of 60 ips.

Figure 6.14   FM tape recorder used for sound recording

Most FM tape recorders have several recording channels. Small battery-powered systems usually have two or four channels. Larger instruments such as that shown in figure 6.14 can have up to 14 recording channels.

Taped records are useful when it is necessary to shift the frequency range of a recording. Low-frequency records, say from vibration measurements, can be shifted into the frequency range of normal audio frequency analysis equipment by replaying the tape at a higher speed than was used for recording. Alternatively, high-speed events such as transients may be slowed down to allow the waveforms to be examined.

6.9   ANALOGUE SPECTRUM ANALYSIS

In sound measurement we are often interested in the amplitude of the signal at a particular frequency or within a certain frequency band, rather than the overall linear or A-weighted sound level. As we saw in chapters 4 and 5, the instruments used for studying frequency domain behaviour are called spectrum analysers. Many spectrum analysers are digital and use the Fourier Transform. One way of visualising the forward Fourier Transform is to see it as a contiguous set of fixed-width narrowband filters, one for each point in the frequency domain. By controlling the sample rate and the number of data points involved in each transform, the bandwidth of these 'filters' may be varied as shown in table 5.1. Usually the bandwidth is a few Hertz at most, and it can be a very small fraction of a Hertz if specialised techniques such as zoom FFT are used [5].

However, it is more useful in estimating the loudness, annoyance or subjective response to a noise to know the mean amplitude within frequency bands $\frac{1}{3}$ or 1 octave wide. This is because the human ear does not respond like a digital spectrum analyser, separating the frequencies present in a sound into many narrow bands. Instead it classifies the spectral components of a noise into frequency bands roughly $\frac{1}{3}$ of an octave wide [1]. To emulate this form of frequency analysis in a spectrum analyser, it is usual to use an arrangement of analogue filters rather than a digital process.

Two main types of frequency analysis are used in spectrum analysers: constant percentage bandwidth and constant bandwidth. Digital (FFT) systems are constant bandwidth, in that the bandwidth of each of the imaginary filters described above is a fixed number of Hertz which is correlated to the centre frequency.

Analogue spectrum analysis systems are usually of the constant percentage bandwidth type. This means that each filter has a bandwidth which is a constant percentage of its centre frequency. The most common bandwidths are $\frac{1}{3}$ octave and 1 octave. Clearly, as the centre frequency increases the bandwidth of the filter in Hertz also increases.

It should be noted that the standard A-, B-, C- and D-weighting curves are also bandpass filters. They have a relatively wide bandwidth, and the roll-off rates at the ends of the spectrum are designed to represent the responses of the human ear.

There are three kinds of analogue filter spectrum analyser. They can be subdivided into two classes: parallel or real-time analysers, which are used for transients and non-stationary signals, and sequential analysers used for continuous stationary noise. There are two kinds of sequential analyser: contiguous-filter systems and swept-filter analysers.

Contiguous-filter analysers

Figure 6.15 shows a block diagram of a typical contiguous filter analyser. The signal is applied in parallel to a bank of bandpass filters which are adjacent in the frequency domain. A detector is applied sequentially to

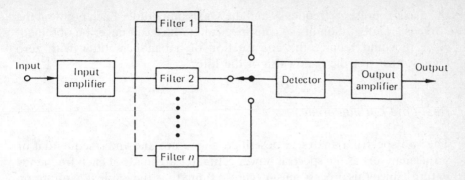

Figure 6.15   Block diagram of contiguous filter spectrum analyser

each filter output, and thus successively measures the power within each frequency band. With this arrangement it is not necessary to wait for the filter response time after switching, but only for that of the detector, which is usually very fast. The power output through each filter is usually plotted on a chart recorder in the form of a histogram. This approach is not normally used for bandwidths below $\frac{1}{3}$ of an octave, since the number of filters needed to obtain higher resolution is prohibitively expensive.

The problem with this sort of spectrum analyser is that it is relatively inflexible. Usually only two resolutions are available, $\frac{1}{3}$ and 1 octave, and the frequency range that can be covered is determined by the characteristics of the filters used and is not usually adjustable.

Swept-filter analysers

For narrow-band analysis it is more common to use a single filter with a tunable centre frequency as illustrated by the block diagram of figure 6.16. The contiguous-filter analyser (described above) provides a number of spectral power estimates at points centred within discrete frequency bands. The spectrum obtained from a swept-filter analyser, however, is continuous. The output is obtained in the form of a chart recorder graph, where

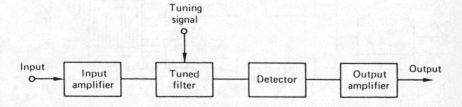

Figure 6.16   Block diagram of swept-filter spectrum analyser

the x-axis represents frequency and the y-axis amplitude. Each point on the curve is an integration of the true spectrum (which can never be obtained, since it would require the construction of a bandpass filter with zero bandwidth) over the bandwidth of the filter.

Real-time parallel analysers

The two spectrum analysers described above are known as sequential or serial analysers as the spectral power estimates are made at each frequency in turn. Since this process must occupy a finite time the implicit assumption is made that the signal is stationary [6]. This means essentially that its spectral content does not vary with time. The formal definition of stationarity (and the basis of testing for stationarity) is that a signal is stationary if

(a) the average value of a signal is constant, and
(b) the average value of the product of two samples of the signal taken at different times (that is, the autocorrelation, see reference [6]) is constant

If a signal is not stationary, its spectrum may change between the first and last filtering operations carried out by a sequential analyser. Sometimes it is possible to force a signal to be stationary, by recording it on a loop of tape which can then be played back continuously. However this procedure is of doubtful validity and can give misleading results, since it imposes an artificial periodicity on the time domain data.

Real-time analysers obtain the whole spectrum in parallel almost instantaneously from the same section of signal. The block diagram of a typical system is shown in figure 6.17. These instruments can not only display the spectral content of non-stationary signals, but are much faster

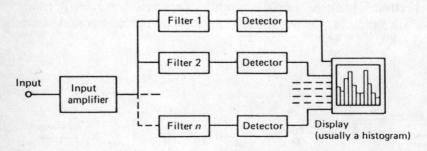

Figure 6.17   Real-time parallel filter spectrum analyser

in operation than sequential analysers since there it is not necessary to wait for the electronic circuits to 'settle'. The most direct way of implementing a real-time analyser is to apply the signal in parallel to a bank of contiguous filter/detector channels as shown in figure 6.13. The speed with which the results are produced and updated makes it essential to have some form of video display. The resolution of this type of spectrum analyser is not normally greater than $\frac{1}{3}$ octave, because of the cost and complexity involved in constructing large numbers of filter/detector units.

REFERENCES

[1] J. R. Hassal and K. Zaveri, *Acoustic Noise Measurements*, 5th edn, Bruel & Kjaer (1988).
[2] Kinsler, Frey, Coppens and Sanders, *Fundamentals of Acoustics*, 3rd edn, Wiley (1982).
[3] F. J. Fahy, *Sound Intensity*, Elsevier (1989).
[4] J. D. Turner, *Instrumentation for Engineers*, Macmillan (1988).
[5] R. B. Randall, *Frequency Analysis*, 3rd edn, Bruel & Kjaer (1987).
[6] J. Bendat and A. Piersol, *Engineering Applications of Correlation and Spectral Analysis,* Wiley (1980).
[7] D. Brook and R. J. Wynne, *Signal Processing Principles and Applications*, Edward Arnold (1988).

7. Reflection and transmission: sound in confined spaces

"Noise is manufactured in the city just as goods are manufactured. The city is the place where noise is kept in stock, completely detached from the object from which it came."
W. H. Auden, *A Certain World*

7.1 INTRODUCTION

Many noise control problems arise indoors where the reflective nature of the walls affects the build up of noise energy from a source. Energy is also transmitted through walls from one room to another. The mechanics of reflection and transmission therefore has a strong influence on the sound levels to which we are subjected inside buildings.

This chapter starts by explaining the classical theory for the reflection of an harmonic acoustic wave arriving at normal incidence at a boundary between two materials. The result of this theory sheds some light on the nature of reflections at walls, and also has special applications in ultrasonic technology. The absorption of energy in reflections is next considered (the classical theory does not account for this). Sound absorption is the mechanism responsible for limiting sound levels in rooms. The noise engineer therefore needs to have quantitative information about absorption if he is to predict sound levels. Transmission of sound energy through walls forms the final part of this chapter.

The material described in this chapter has obvious relevance to sound in buildings. However, it has applications in other situations too, such as in the design of acoustic enclosures for machines.

7.2 REFLECTION OF PLANE WAVES AT A BOUNDARY

> *"The other Messenger's called Hatta. I must have*
> *two you know, to come and to go. One to come*
> *and one to go."*
> Lewis Carroll, *Through the Looking Glass*

The reflection and transmission of plane waves at a plane boundary
between two media of different acoustic impedance is now examined.
These waves are one-dimensional and so reference can be made to some of
the results of chapter 1. The theory does not account for oblique incidence
(that is, incidence angles other than 90°) nor does it allow for any energy
loss at reflection. As regards the walls in rooms, the theory is only directly
applicable to the reflected wave since a transmitted wave has to cross two
boundaries. This will be dealt with later in the chapter.

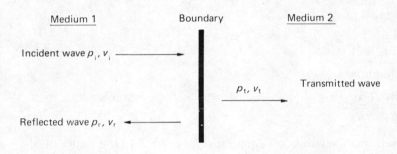

Figure 7.1

In figure 7.1 the incoming (incident) wave from the left falls upon the
boundary. It is then split into two components, one part of the energy
being transmitted and the other reflected. In the following set of equations,
the pressure in each wave is defined as a sinusoid of given amplitude A and
circular frequency ω, in the form of equation 1.6. In this definition, use is
made of the wavenumber k, as in chapter 3, though in fact it is a mere
convenience and is not involved in the results of the theory. Distance x is
measured from the boundary, positive to the right. The equations also
include the impedance relationships (equations 1.14 and 1.15) appropriate
to each wave. R is the specific acoustic impedance. Note that in the third of
the following equations the appropriate wavenumber is k_2 as the transmitted wave is in a different medium.

$$p_i = A_i \sin(\omega t - k_1 x) = v_i R_1$$

$$p_r = A_r \sin(\omega t + k_1 x) = -v_r R_1$$

$$p_t = A_t \sin(\omega t - k_2 x) = v_t R_2 \tag{7.1}$$

In medium (1), to the left, the actual pressures (p), displacements, and velocities (v, positive to the right) are a linear superposition of the quantities associated with each of the two waves. The conditions at the interface are that the total pressure and the total velocity are the same in both media at the boundary. Thus

$$p_i + p_r = p_t$$

$$v_i + v_r = v_t \tag{7.2}$$

At the boundary ($x = 0$) these give the

pressure condition

$$A_i + A_r = A_t \tag{7.3}$$

velocity condition

$$\frac{A_i}{R_1} - \frac{A_r}{R_1} = \frac{A_t}{R_2} \tag{7.4}$$

By algebraic manipulation of equations 7.3 and 7.4, the ratio of the amplitudes of the transmitted and incident waves is found:

$$\frac{p_t}{p_i} = \frac{A_t}{A_i} = \frac{2R_2}{R_1 + R_2} \tag{7.5}$$

and the ratio of amplitudes of the reflected and incident waves is

$$\frac{p_r}{p_i} = \frac{A_r}{A_i} = \frac{R_2 - R_1}{R_2 + R_1} \tag{7.6}$$

Equation 7.6 shows that if $R_2 \gg R_1$, as it would be at the surface of a wall in a room, then $p_r/p_i = 1$ and the TOTAL pressure at the surface will be $p_i + p_r = 2p_i$. This effect is known as *pressure doubling*. In chapter 3, the importance of this effect on directivity was described. Two other examples of importance are the design of buildings to resist blast or shock waves and the correct calibration of microphones.

The velocity ratios can also easily be derived

$$\frac{v_t}{v_i} = \frac{2R_1}{R_1 + R_2} \tag{7.7}$$

and

$$\frac{v_r}{v_i} = \frac{R_1 - R_2}{R_1 + R_2} \tag{7.8}$$

The energy intensity in a given wave is the product of pressure and velocity and so the energy ratios in the wave are

$$\frac{I_r}{I_i} = \frac{v_r p_r}{v_i p_i} = \alpha_r = \frac{(R_2 - R_1)^2}{(R_2 + R_1)^2} \tag{7.9}$$

and

$$\frac{I_t}{I_i} = \frac{v_t p_t}{v_i p_i} = \alpha_t = \frac{4R_1 R_2}{(R_1 + R_2)^2} \tag{7.10}$$

α_r and α_t are called the reflection and transmission coefficients respectively. Clearly, the transmission coefficient is greatest when $R_1 = R_2$. This corresponds exactly with the classic impedance matching of electric circuits to obtain maximum power transfer. Under this condition:

$$\alpha_r = 0$$
$$\alpha_t = 1$$

On the other hand, if R_1 is very different from R_2, as it would be for reflection at a wall, then α_t becomes small and α_r becomes close to unity. This is the main reason why little energy is transmitted from one room to another. The principle of conservation of energy shows that

$$\alpha_r + \alpha_t = 1 \tag{7.11}$$

This may be confirmed by substitution of equations 7.9 and 7.10 into 7.11.

In most problems associated with noise control it is desirable to achieve the maximum mismatch of impedance in order to prevent transmission of noise. However, in some acoustic applications, such as in the design of ultrasonic transducers for injecting noise energy into a fluid, matching of impedance is desirable so as to maximise the energy transfer. A further example is the design of sonar transducers for use at sea. It can be shown that if a third medium, of impedance R', is placed between the two media (1) and (2) in figure 7.1:

(a) if the thickness is equal to an integral number of half-wavelengths then α_t is given by equation 7.10;
(b) if the thickness is equal to an odd number of quarter wavelengths then α_t is given by

$$\alpha_t = \frac{4R_1R_2}{\left(R' + \dfrac{R_1R_2}{R'}\right)^2} \tag{7.12}$$

Equation (7.12) is significant because if $R' = \sqrt{(R_1R_2)}$, $\alpha_t = 1$. Thus if there is an impedance mismatch between two areas, a component can be designed which will fit between the two media and will provide a matching device for a given frequency. An appropriate material cannot always be found, of course. Such a matching component is called, by analogy with electrical circuits, a transformer. If it is only one quarter wavelength long it is called a 'quarter-wave transformer'.

Generally, there is extremely poor acoustic coupling between liquids or solids and gases because of the impedance mismatch, see table 1.1 for relevant data. This is why it is difficult to hear sounds originating in the air when underwater or vice versa. However, sonar apparatus transmits sound underwater which can travel very large distances, partly because the energy in the waves never gets into the atmosphere and partly because it is effectively propagated in two dimensions instead of three.

The theory outlined above is very similar to the theories for reflection at plane surfaces for other types of wave such as in optics and radio transmission. This can lead to the use of optical analogies for the study of the acoustics of rooms and auditoria. For sound which is obliquely incident upon a plane boundary there are refraction properties exactly the same as those for light waves. The same law, Snell's law, is obeyed. Total internal reflection can also occur with acoustic waves. The phenomenon of diffraction and scattering is identical in character to the similar phenomenon for light and it has already been seen (chapter 3) that it is possible to obtain acoustic 'shadows' when the wavelength of sound is much smaller than the obstacle.

Example

A nickel magnetostrictive transducer is to be used as the driver for an ultrasonic cleaning bath operating at 22 kHz in water. Using equation 7.12, design the best available and practicable quarter-wave transformer from the data for materials given in table 1.1. Work out the transmission coefficient and the length of the transformer.

Solution

Under ideal conditions a material is required with a characteristic impedance which is the geometric mean of the values for nickel and water. This is

$$R' = \sqrt{[(4.4 \times 10^7) \times (1.5 \times 10^6)]} = 8.1 \times 10^6$$

From table 1.1, concrete appears to be the appropriate material. However, it is quite unsuitable in practice because of the difficulty of attaching it securely and its inability to carry tensile stresses. The nearest suitable material is glass for which $R = 1.2 \times 10^7$. The transmission coefficient may now be calculated using equation 7.12 with

$$R_1 = 4.4 \times 10^7$$
$$R_2 = 1.5 \times 10^7$$
$$R' = 1.2 \times 10^7$$

This gives

$$\alpha_t = 0.56$$

The speed of sound in glass is 5000 m/s so the appropriate length (a quarter wavelength) at 22 kHz is

$$\left(\frac{5000}{22000} \times \frac{1}{4}\right) = 57 \text{ mm}$$

7.3 EXPANSION CHAMBER SILENCERS

It is appropriate at this point to give a brief account of expansion chamber silencers because their performance is based on a theory closely similar to that given above. This type of silencer is usually associated with motor car exhausts but it is also used for high flow gas vents of all kinds, in fan ducting and even in situations where there is no gas flow at all, such as entry ports to noisy enclosures. Such devices are classified as reactive (as opposed to dissipative) mufflers. The theory for complete systems incorporating these elements is generally very complex and electrical circuit analogies are often used. A full account of this theory is contained in

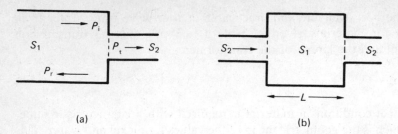

(a)   (b)

Figure 7.2

reference [1]. Here, the properties of the basic expansion chamber, as used on engine exhausts, will be examined.

Figure 7.2(a) shows a change of section in a round pipe of infinite extent in both directions with sound incident from the left. The cross-sectional area of the pipe is denoted by S. For the purposes of this theory, the acoustic impedance will be defined in a slightly different way as

$$R_v = \frac{\text{pressure}}{\text{volume velocity}} \tag{7.13}$$

This is, of course, directly related to the specific acoustic impedance because

Volume velocity = S × Particle velocity

Hence

$$R_v = \frac{R}{S} = \frac{\rho c}{S}$$

By contrast with the previous theory, the gas is assumed to be the same on each side of the discontinuity but the areas are different. The conditions at the discontinuity are that total pressure and the total *volume* velocity are the same. Hence, by comparison with the previous theory, R_v can be used in place of R. As a result, the transmission coefficient (see equation 7.10) is

$$\alpha_t = \frac{4R_{v1}R_{v2}}{(R_{v1} + R_{v2})^2} = \frac{4S_1S_2}{(S_1 + S_2)^2} \tag{7.14}$$

If $S_1 = nS_2$:

$$\alpha_t = 4n/(n+1)^2 \tag{7.15}$$

If n is large, this approximates to

$$\alpha_t = 4/n \tag{7.16}$$

The general conclusion is that transmission is reduced as the area ratio n increases. The same result is obtained for transmission from a small to a large pipe.

Consider next an expansion chamber as in figure 7.2(b). The analysis is now complicated by the double reflections which take place. For the chamber to work satisfactorily its length needs to be small compared with the sound wavelength otherwise there may be standing waves which will reduce its effectiveness. However, it must not be too short or the sound will pass straight through. The ideal is when the length is equal to a quarter-wavelength. The characteristics are best described by the data shown in figure 7.3.

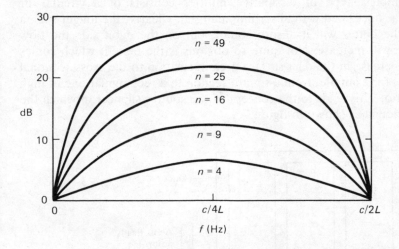

Figure 7.3   *Attenuation for a single expansion chamber of area ratio* **n**, *see figure 7.2(b)*

In practical cases, the pipes are not infinitely long and the effect of the tail pipe and the impedance at the source end both have significant effects. The flow velocity also has a small effect. The data in figure 7.3 should therefore only be taken as a guide. Many modern vehicles have multiple expansion boxes and absorbing material is usually installed inside the expansion chambers. An important consideration in practice is that the silencing system exerts considerable back pressure on the engine when it is working effectively. This will reduce the efficiency of the engine itself because it cannot then dispose of the exhaust gases so readily.

7.4 DISSIPATIVE SILENCERS

We have mentioned above that acoustically absorbing material may line the walls of expansion chamber silencers. This type of material can also be incorporated in other ways to reduce the passage of sound in piped or ducted systems. The prime mechanism here is to absorb sound energy rather than to reflect it back the way it came. Such devices are widely used where there is a steady gas flow, such as in gas turbine exhausts or in air-conditioning systems. The absorbing materials are usually fibrous or porous and for reasons described in the next section this leads to absorption primarily at mid- to high-frequencies. Such devices are often called 'mufflers'.

The simplest type of dissipative muffler consists of a circular or rectangular duct lined with absorbing material. Clearly, the longer such a duct is, the better will it absorb sound, but on the debit side the flow resistance will increase. Not quite so obvious is the benefit which comes from an increase in the duct surface area in relation to the cross-sectional geometry, for example, from a square section to a rectangular one of high aspect ratio. This leads to the concept of the splitter silencer in which the cross-section is as shown in figure 7.4.

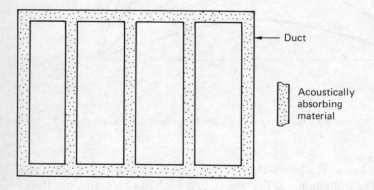

Figure 7.4 Cross-section of a splitter

However, this will again have the undesirable consequence of increasing the flow resistance. If a splitter design is installed in a duct of uniform size it also has the undesirable effect of increasing the flow velocity which, of itself, will increase the noise generation due to shear and turbulence. It is therefore desirable to have not only long sections of muffler but also ones of increased external dimensions. This leads to undesirable increase in bulk and cost for the system. In design therefore there is considerable scope for ingenuity in resolving conflicts which arise between, on the one hand, better acoustic performance and, on the other, cost, bulk, increased back pressure and secondary noise generation. Reference [1] devotes a whole chapter to the design of mufflers and the interested reader is referred to this for more details.

7.5   ACOUSTIC ABSORPTION AT SURFACES IN ROOMS

The simple theory given earlier in this chapter indicates near perfect reflection at room surfaces. If there were perfect reflection then the sound energy from the source would continuously build-up in a closed room and the intensity would become very large: what is more, the sound would continue unabated when the source was turned off. Neither of these things happens. The reason is that sound energy is absorbed and converted to heat or lost in other ways. To a limited extent, the sound energy may be absorbed as it passes through the air, see figure 3.12. However this is usually very small except at high frequencies in very large rooms, and so this will be disgregarded. Sound energy may also be lost from the room by passing through open gaps such as doorways or open windows. The major loss of sound energy occurs upon reflection at surfaces, including not only the fixed surfaces in a room but also mobile surfaces such as the occupants. The mechanism of energy loss is usually one of viscous dissipation as the air oscillates at or near surfaces. The texture of the surface is therefore very important. Fibrous surfaces, such as curtain fabrics or glass fibre tiles, are very absorptive whereas hard concrete surfaces are highly reflective. The quality of a surface in this respect is quantified by the ABSORPTION COEFFICIENT (α; not to be confused with reflection and transmission coefficients) which can vary from near zero, for a hard reflective surface such as concrete, to near unity, for a fine fibrous surface such as a glass fibre tile. Absorption coefficients for a number of typical room surfaces are shown in table 7.1. It is seen that values of α vary considerably with frequency. The absorption coefficient can also vary with angle of incidence. In a room, sound impinges on surfaces from all directions and an average absorption coefficient is then required. This average value can be measured for a given material in a specially constructed reverberant room facility. However, it is more usual to use a standing wave tube to measure the absorption coefficient because it provides a much simpler and cheaper test. This laboratory instrument consists of a long narrow tube with a loudspeaker as a source of sound at one end and a sample of the relevant material at the other. The absorption coefficient of the material can be deduced from the nature of the standing wave which is set up in the tube at various frequencies. However, the value which is then obtained is for normal incidence only. In practice, the two values are relatively close for most practical materials.

Table 7.1 Approximate values of absorption coefficients for some typical building materials averaged in octave bands.

Material	Octave band centre frequency (Hz)						
	62.5	*125*	*250*	*500*	*1000*	*2000*	*4000*
Open window*	1.00	1.00	1.00	1.00	1.00	1.00	1.00
Glass**: 3 mm sheet	—	0.3	—	0.1	—	0.05	—
6 mm plate	—	0.1	—	0.04	—	0.02	—
Plastered brickwork (concrete similar)	0.04	0.03	0.03	0.03	0.03	0.04	0.04
Lightweight blocks (not plastered)	0.1	0.2	0.3	0.6	0.6	0.5	0.5
Suspended plasterboard ceiling***	0.1	0.3	0.15	0.1	0.05	0.04	0.05
Plastic floor covering on concrete base	0.05	0.02	0.04	0.05	0.05	0.1	0.05
Floor boards on joists over a gap	0.1	0.3	0.2	0.1	0.1	0.1	0.05
Pile carpet on underfelt on concrete base	0.05	0.07	0.3	0.5	0.5	0.6	0.7
Acoustic tiles on solid wall†	0.15	0.4	0.65	0.85	0.9	0.65	0.7

Notes

  * Unity absorption coefficient at low frequency only if the opening is very large (that is, greater than two or three wavelengths).

 ** High absorption at low frequency because the window acts as a panel absorber. This frequency will depend on window dimensions.

*** Acting like a panel absorber.

  † Acoustic tiles vary considerably. Some absorb partly as a Helmholtz resonator. If mounted on battens over a gap they will provide additional low-frequency absorption.

The energy can be lost by three mechanisms:

(a) viscous flow losses in fluffy or textured surfaces, as already mentioned;
(b) mechanical resonance effects at room surfaces;
(c) transmission through holes, such as open windows.

For viscous flow losses to be effective the mean acoustic velocity and the velocity shear rate at the fibre surfaces must both be maximised. If this is to be achieved by means of a specially designed tile, then the first requires either a maximum depth of fibrous material or a gap separating it from the reflecting wall to which it is attached or both. This is because the acoustic velocity is zero *at* the reflecting wall. The maximum shear rate is achieved by using fine fibres, relatively close together. However, there is an optimum fibre packing. If the fibres are packed too close together sound is reflected from them, rather than passing through, and the tile will be less effective. Such tiles must have a porous wearing surface on the outside and this has to be designed carefully to provide a route for the ingress of sound. Care must be taken too in decorating such tiles: paint can block the pores in the surface. Materials such as carpets and clothes do not fulfil the velocity requirement except at high frequencies when the wavelength is shorter. This is observed in the corresponding entry in table 7.1. Generally, fibrous and textured surfaces have poor absorption at low frequencies but very good absorption at high frequencies.

Mechanical resonance absorbers are only effective at frequencies close to their natural resonance frequencies. This provides the opportunity to complement the relatively poor low-frequency performance of fibrous absorbers. Therefore absorbers of this type are often designed to have natural frequencies in the range 50–200 Hz. Mechanical resonance absorption, whether deliberately provided or otherwise, is of two types. The first is where structural or decorative components of the room are in resonance with the sound. Such components might be, for example, panelling, window panes, floor panels, etc. They will absorb energy at their resonance frequency by means of internal material damping.

Artificial absorbers of this kind usually consist of closed boxes on the front of which is attached a limp panel. The panel, often of hardboard, is usually designed to have relatively high internal damping by gluing roofing felt or a similar high loss material to its inside surface. The formula for calculating its natural frequency is

$$f_1 = \frac{c}{2\pi} \sqrt{\left(\frac{\rho}{md}\right)} \tag{7.17}$$

where

c = speed of sound
ρ = density of air
m = mass per unit area of the panel
d = depth of the box behind the panel.

The second type is often called a Helmholtz resonator and consists of an air resonator like a bottle. A resonance note can be obtained from a bottle by blowing across its open end. The same bottle will absorb sound strongly at this frequency. The explanation is that the plug of air in the neck of the bottle acts as a mass vibrating on the spring provided by the air volume inside the bottle. An approximate formula for calculating its natural frequency is

$$f_1 = \frac{c}{2\pi} \left\{ \frac{\pi a^2}{V(L + \frac{\pi a}{2})} \right\}^{\frac{1}{2}} \tag{7.18}$$

where

c = speed of sound
a = neck radius
L = neck length
V = bottle volume.

It is unlikely that you will see a range of different size bottles lying around a room for this purpose. The Helmholtz resonator does not, however, need to be in the form of a bottle. Figure 7.5 shows a cross-section of a perforated acoustic tile which behaves in this way, together with its absorption characteristic. A different form is the slotted tile which, while working in a similar way, is said to be more aesthetically attractive. If, in figure 7.5, the perforated sheet is thin (usually metal) compared with the perforation size then, for $L = 0$, equation 7.18 reduces to

$$f_1 = \frac{c}{2\pi} \left(\frac{2a}{V} \right)^{\frac{1}{2}}$$

In passing, it is worth mentioning that Helmholtz resonators are used in many noise-reducing applications apart from their use as frequency selective absorbers in rooms. A typical example is in fan ducting where the noise source has a predominant pure tone due to the fan blade passage frequency. The resonator, in this case very often bottle-shaped but made of metal, is attached to the wall of the ducting at a suitable point.

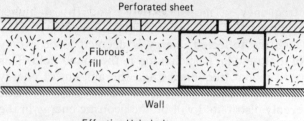

Perforated sheet

Fibrous fill

Wall

Effective Helmholtz resonator
shown in heavy line

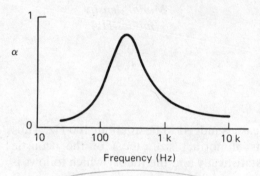

Frequency (Hz)

Figure 7.5

7.6 SOUND IN ENCLOSED SPACES: ROOM ACOUSTICS

The overall sound field in a room is composed of two components, the direct sound from the source and sound reflected from the walls. Unless we are close to the source, the reflected or reverberant sound energy tends to dominate. The room itself has standing wave properties rather like an organ pipe, but much more complex because the standing waves occur in three dimensions. At low frequencies these modes of acoustical vibration are well separated on the frequency scale and so they can distinctly affect the acoustic feel of a room. Higher up the frequency scale, the modal frequencies are much closer together and so a pure tone will significantly excite several modes together. As a result, sound energy is much more evenly distributed throughout the room and simple averaging or statistical methods can be used, such as the thermal analogy mentioned later in this chapter. This approach is also the basis for the widely used Statistical Energy Analysis (SEA) methods for the analysis of sound and vibration problems which have been the subject of much research in recent years. Thus, in the analysis of room acoustics it appears to be necessary to use two methods: (a) at low frequency, a modal analysis, and (b) at higher frequencies, a statistical approach. The transition between the two is at the frequency value for which the reflected energy is deemed to be evenly distributed as a result of multi-modal participation in the sound field. This depends on the density of modal frequencies and the amount of energy absorption in the room: the greater the modal density and acoustic absorption, the more evenly distributed will be the acoustic energy in the room. As an example, for a room measuring $10 \times 5 \times 3$ metres (a large living room) the modal density is as follows:

Frequency (Hz)	Modal density (modes/Hz)
62.5	0.2
125	0.9
250	3.5
500	15

The transition frequency in this case is judged to be at about 100 Hz. Thus it can be seen that, for rooms of normal size, most of the acoustic frequency range can be treated statistically and the theory which follows is based on this assumption. However, this is an approximation and it sometimes leads to odd results, such as absorption coefficients greater than one. In addition, at low frequencies it can sometimes be vitally necessary to take a modal view.

Returning to the effect of acoustic absorption at the surfaces of rooms; can we say that there are ideal values? To answer this another question must be asked: what are the ideal room characteristics? To answer this, yet another question has to be asked: how is a room characterised? The answer to this is concerned with how reverberant (echoey) a room is. We need to know how quickly the sound dies away when a source is turned off. The speed with which this happens is described by what is termed the Reverberation Time (**RT**). This is defined as the time it takes for the sound pressure level to drop by 60 dB. It clearly depends on the amount of absorption in the room and also on the dimensions of the room. Rooms with a larger volume may be expected to have longer **RT**s because, on average, there are less reflections per unit time. Typical values are 0.5 second for a small room, ten seconds for a cathedral. For the optimum of listening conditions in a room it is required that the **RT** be neither too short nor too long. If there is too much absorption then the room feels dead and only direct sound reaches the listener. The steady-state level will then be lower than if echoes were present. As a result there may be a problem in hearing a speaker. On the other hand, if the **RT** is too long the problem is not one of being able to hear but rather of being able to understand. The multiple echoes in the room tend to mask the direct sound and render it unintelligible. In general, the optimum **RT** is different for different conditions. The ideal **RT** for speech is generally less than for music. For speech, clarity is required, whereas for music some echo enriches the tone. Perhaps that is why we so much enjoy singing in the bath. Optimum values for different types of room are shown in figure 7.6. It is interesting to speculate that plainsong chanting developed in the Sixth Century monasteries in response to the long **RT**s of the abbey buildings. On the other hand, modern dissonant music such as heavy metal rock is best performed in the open air at rock or pop festivals (some might say that down a deep coal mine would be better).

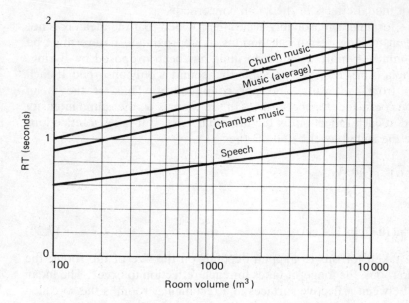

Figure 7.6   Optimum Reverberation Times

The **RT** can be calculated using the simple approximate formula proposed by Sabine:

$$\textbf{RT} = \frac{\textbf{0.16 } V}{A} \tag{7.19}$$

where

V = room volume in m^3
A = total absorption in m^2.

$$A = \sum_n S_1\alpha_1 + S_2\alpha_2 + ... + S_n\alpha_n$$

with S_n = Area of material n in m^2

α_n = the corresponding absorption coefficient.

Note that this form of Sabine's formula requires the various parameters to be in SI units. The **RT** is usually measured using broad band noise but it is possible to measure it for narrow band sounds. The broad band value will be some sort of average of narrow band values. The broad band value is usually quoted over the frequency range 250 Hz to 4 kHz. Single frequency values may have considerable variability over short frequency intervals because of modal effects in the room concerned.

Sabine's formula is reasonably accurate provided that the mean α is not high, but in making design calculations its approximate nature must be borne in mind. A more accurate formula has been proposed by Eyring. This formula is based on the principle that sound is only absorbed at each reflection from the walls, and not continuously. Thus, if the mean absorption coefficient for the surfaces of the room is $\overline{\alpha}$, the sound intensity is reduced at each reflection by the factor $(1 - \overline{\alpha})$. If it takes n reflections to reduce the sound level by 60 dB then

$$(1 - \overline{\alpha})^n = 10^{-6}$$

and hence

$$n = \ln(10^{-6})/\ln(1 - \overline{\alpha}) \tag{7.20}$$

The time taken for this to happen depends on the size of the room; the bigger the room, the longer it takes for each reflection to occur. The mean distance between reflective surfaces in a rectangular room is the so-called 'mean free path' and this is given by the formula

$$h = 4V/S \tag{7.21}$$

Hence if the time taken for n reflections is the Reverberation Time (**RT**) then

$$\mathbf{RT} = nh/c = 4nV/cS \tag{7.22}$$

Substituting equation 7.20 in 7.22:

$$\mathbf{RT} = \frac{4V}{cS} \frac{\ln 10^{-6}}{\ln(1 - \overline{\alpha})}$$

or

$$RT = \frac{0.16V}{-S\ln(1 - \overline{\alpha})} \qquad \text{(for } c = 330 \text{ m/s)} \qquad (7.23)$$

This is Eyring's formula. If α is small (0.1 or less), then

$$\ln(1 - \overline{\alpha}) \approx -\overline{\alpha}$$

and the formula becomes the same as Sabine's formula. At high frequency in large rooms, absorption in the air may need to be taken into account.

7.7   RT CALCULATIONS IN PROBLEM SOLVING

The general technique is to aim at a particular **RT** for a room in accordance with established standards of good practice. For rooms designed for speech or for music, the standards are given in figure 7.5. For more unusual rooms or spaces, there is a great deal of information in reference [3]. In some circumstances, and perhaps especially in industrial situations, there is a need to curb reverberation at some particular frequency either because the room is unduly resonant or because there is a source of near harmonic noise. In such cases, use can be made of tuned or narrow band absorbers of the kind described earlier in this chapter.

7.8   CASE STUDY: ACOUSTIC TREATMENT OF READING CENTRAL SWIMMING BATHS

Swimming baths present some special difficulties in selecting acoustic treatment. Generally, tiles made with natural fibres are not satisfactory because of the high humidity and the consequent risk of fungal rot. The high level of chlorination in the water leads to an atmosphere which is especially corrosive of steel and also of some other metals, so metal panels and fixtures have to be chosen with great care. The commercial tiling X, which was chosen for this application, is a perforated aluminium panel with fibreglass backing, which meets these difficulties.

Some years ago, as a result of a structural failure in the ceiling of the Reading baths, the local authority decided to fit new ceiling material which would also provide a reduction in acoustic reverberation. Indoor swimming pools are notoriously noisy places because the water surface and glazed tile walls provide surfaces which are highly reflective. As a result there is a considerable safety hazard if a swimmer gets into difficulty and either his cries or the steward's whistle go unheard. In reference [3], para. 706, it is recommended that the **RT** should not be greater than 2.5 seconds for swimming baths. The following summary shows, in brief, how this was achieved by the consultant concerned. Use is made of equation 7.19. Basic data for the baths is

Room volume = 6000 m^3
Ceiling area   = 800 m^2

Initial measurements were made without a ceiling fitted using octave-band noise as a source. Once the sound level had reached a steady value, the source was disconnected and the decay of sound level measured. The results are shown in columns 1 and 2 of the following table, and column 3 shows the derived value of the total absorption A using equation 7.19. The **RT**-values in column 2 are very high compared with the standard.

Measurements made before treatment (octave bands)

1 Frequency (Hz)	2 RT (seconds)	3 Existing absorption (m^2)
125	4.4	220
250	5.6	170
500	6.4	150
1000	5.7	168
2000	4.1	234
4000	2.9	331

Sabine's formula indicates that the total absorption required is

$$A = \frac{0.16V}{RT} = \frac{0.16 \times 6000}{2.5} = 384 \text{ m}^2$$

In the tabular calculation which follows, the aim is to determine the required area of acoustic tiling X (column 4) in order to achieve an **RT** of 2.5 seconds at each frequency. The maximum value then defines the minimum area to be fitted. Column 2 shows the additional absorption required to bring the existing values up to 384 m^2. Column 3 shows the quoted manufacturer's value for the panel absorption coefficient. Calculations using values in columns 2 and 3 give the required areas of treatment in column 4.

Calculations for treatment

1	2	3	4
			Tiling X
Frequency (Hz)	*Additional minimum absorption (m^2)*	*Alleged absorption coefficient*	*Area required (m^2)*
125	164	0.62	265
250	214	0.87	246
500	234	0.82	285
1000	216	0.95	227
2000	150	0.92	163
4000	53	0.95	56

The conclusion of this calculation is that it is necessary to fit at least 285 m^2 of tiling. In the event it was decided to fit 360 m^2 to the 800 m^2 ceiling, and the following table shows the resulting prediction for **RT**s and the corresponding subsequent measurements.

Prediction and check measurements

Frequency (Hz)	*Predicted RT (seconds)*	*Measured RT (seconds)*
125	2.17	2.85
250	1.99	3.05
500	2.16	2.30
1000	1.88	2.30
2000	1.70	2.25
4000	1.43	1.80

It is seen that the measured **RT** is greater than the predicted at all frequencies, and is particularly erroneous at low frequency. There are three points to be made here: (1) manufacturers tend to quote optimistic figures, (2) low-frequency values are particularly hard to trust because they usually depend on panel resonance effects, and (3) Sabine's formula is only approximate. As it happens, the client was very happy with the result obtained, even though the standard value was not entirely achieved.

7.9   ESTIMATING THE SPL FOR A ROOM OF KNOWN RT, FOR A SOURCE OF GIVEN POWER

The sound output power of a source, (and its directionality if this is required) may be measured by means of tests in a specially constructed anechoic chamber. This is a chamber with perfectly absorbing walls (in principle) which thus simulates an open space. Measurements of sound level are made over a hemispherical surface situated at a suitable distance from the source. This can be a very expensive set of tests to carry out, involving, as it does, the use of costly facilities and expert staff. However, many machine manufacturers quote the results of such tests for their products because this is a vital starting point in the calculation of sound levels under reverberant conditions.

In this section a simple calculation is made on the basis of a single figure sound output power value in Watts. This does not account for spectral variations with frequency, so that in a fuller calculation this effect would have to be taken into account by means of multiple calculations. Under reverberant conditions, directionality of the source is generally of little importance because the radiated sound energy tends to become evenly spread throughout the space because of multiple reflections.

The calculation of sound level is based on an energy/intensity view of the processes at work. The situation is closely analogous to a thermal problem. In figure 7.7 the source of acoustic energy is shown as Q_1 (Watts). Energy is lost from the system because of acoustic absorption, shown as Q_2 (Watts). There is also an amount of acoustic energy E (Joules) which is proportional to p^2 and is contained within the room space. The analogous thermal situation in a room is a hot water radiator source (Q_1), heat loss through walls (Q_2), and thermal energy contained in the room volume (E) which is proportional to air temperature. The analogy may help in understanding the theory which follows. The energy E is assumed to be uniformly distributed throughout the room which has a value V.

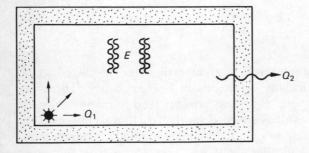

Figure 7.7

In a steady state of acoustic equilibrium, E is fixed and Q_1 balances Q_2. If the source power Q_1 is increased, then a new equilibrium state is reached in which Q_2 also increases until $Q_1 = Q_2$ once again. To achieve this, E must also rise because the absorbed power Q_2 is proportional to the intensity I and thus to E. So

$$Q_2 \propto E \propto \overline{p^2}$$

Hence

$$Q_1 = Q_2 = \overline{kp^2} \qquad (7.24)$$

The energy E is derived in terms of the mean intensity I as

$$E = \frac{IV}{c} \text{ or } E = \frac{\overline{p^2}V}{c^2} \qquad (7.25)$$

This is explained by realising that in a one-dimensional travelling wave an intensity of I W/m^2 fills a volume of c cubic metres in one second with I Joules. This gives an energy density in the air of (I/c) J/m^3. It is reasonable to suppose that the constant k is proportional to the total acoustic absorption and so k should also be directly related to the RT. If k can be found then, for a given Q_1, $\overline{p^2}$ can be determined from equation 7.24 above, and so the aim would be achieved.

To find k, a different situation is now considered. Imagine that the source of energy Q_1 is turned off and there will then be a decay of E due to the continuing loss Q_2. This is precisely the transient state from which the RT was defined. The relationship between Q_2 and E is now

$$Q_2 = \frac{-dE}{dt} \qquad (7.26)$$

Substituting for Q_2 and E from equations 7.24 and 7.25, a differential equation is obtained:

$$\frac{V}{\rho c^2} \cdot \frac{d(\overline{p^2})}{dt} + \overline{kp^2} = 0 \qquad (7.27)$$

The solution to this is an exponential decay of $\overline{p^2}$, thus

$$\overline{p^2}(t) = \overline{p^2}_0 \, e^{-\left(\frac{k\rho c^2 t}{V}\right)}$$

At the moment when Q_1 is turned off, $t = 0$ and $\overline{p^2}(t) = \overline{p_0^2}$. If the **RT** (60 dB drop time) is T then

$$\overline{p^2}(t+T) = \overline{p_0^2}\, e^{-\left(\frac{k\rho c^2(t+T)}{V}\right)}$$

and from the definition of T

$$60 \text{ dB} = 10\log_{10}\left(\frac{\overline{p^2}(t)}{\overline{p^2}(t+T)}\right) = 10\log_{10}e^{\left(\frac{k\rho c^2 T}{V}\right)}$$

giving

$$k = \frac{13.82\ V}{\rho c^2 T} \tag{7.28}$$

Returning to equation 7.24 for the steady state:

$$\overline{p^2} = \frac{Q_1}{k} = \frac{\rho c^2 Q_1 T}{13.82\ V} \tag{7.29}$$

or alternatively for intensity

$$I = \frac{\overline{p^2}}{\rho c} = \frac{c Q_1 T}{13.82\ V} \tag{7.30}$$

Making use of Sabine's formula, equation 7.19, and a value for the speed of sound in air of $c = 345$ m/s, this equation reduces to the simple result

$$I = \frac{4 Q_1}{A} \tag{7.31}$$

It is convenient to quote the source power on a logarithmic scale relative to a standard power of 10^{-12} Watts. This is widely known as the sound power level, **PWL**, and is defined by

$$\text{PWL} = 10\log_{10}\left(\frac{Q_1}{10^{-12}}\right) \tag{7.32}$$

By taking logarithms equation 7.31 can then be expressed as

$$SPL = PWL + 10\log_{10}(4/A) \tag{7.33}$$

Equation 7.33 is now in a suitable form to permit the calculation of the *SPL* in a room for a given known source power provided that the room volume and **RT** are known. It can also be used in reverse to determine the power of a source if the *SPL* is known.

Finally, the limitations of equation 7.33 must be emphasised. The theory is based on the assumption of a perfectly diffuse sound field in the room. This will not be true if the room has focusing qualities (a barrel roof for example), if the measurement position is very close to the source (where direct sound will be correspondingly larger), if other rooms are acoustically coupled to the room in question, or if the distribution of acoustic absorption is very uneven.

A simple modification to equation 7.33 can be made which takes account of the direct sound effect and which can prove more useful than equation 7.33 when trying to estimate the *SPL* at a particular location in a room in the presence of a source of known power. The guiding principle is that the sound at the particular location consists of two parts; first the direct sound, secondly the reverberant sound which must have been reflected from a room surface *at least once*. The reverberant sound therefore arises from a source which is weaker than Q_1 by a factor $(1 - \overline{\alpha})$. Hence, the reverberant intensity, by comparison with equation 7.31, is

$$I_r = \frac{4Q_1(1 - \overline{\alpha})}{S\overline{\alpha}} \tag{7.34}$$

The quantity $S\overline{\alpha}/(1 - \overline{\alpha})$ is called the Room Constant and is generally denoted by the symbol R. To this reverberant intensity must now be added the intensity arising from the direct sound. Including the appropriate Directivity Index *DI* for the relevant point, a distance r from the source, the direct sound intensity is

$$I_d = \frac{Q_1 D}{4\pi r^2} \tag{7.35}$$

Thus the total intensity is

$$I = I_r + I_d = Q_1\left(\frac{D}{4\pi r^2} + \frac{4}{R}\right) \tag{7.36}$$

and it follows that the equation for the *SPL*, which corresponds to equation 7.33, is

$$SPL = PWL + 10 \log_{10}\left(\frac{D}{4\pi r^2} + \frac{4}{R}\right) \tag{7.37}$$

Example

The speech broad band (250–4000 Hz) **RT** for a small lecture theatre of volume 100 m^3 is measured to be 0.5 second. The average **SPL** in the theatre for a certain lecturer is 64 dB. Estimate the mean acoustic power of his voice.

Solution

To answer this question, equation 7.33 is used. First it is necessary to calculate the total absorption in the room from a rearrangement of Sabine's formula:

$$A = \frac{\mathbf{0.16\ V}}{T} = \frac{0.16 \times 100}{0.5} = 32\ \text{m}^2$$

Then

$$64\ \text{dB} = \text{PWL} + 10\log_{10}(4/32) = \text{PWL} -9\ \text{dB}$$

or

$$\text{PWL} = 73\ \text{dB} = 10\log_{10}(Q_1/10^{-12})$$

Hence

$$Q_1 = 2 \times 10^{-5}\ \text{Watts}$$

This calculation shows how little power is required to produce a strong sound in a small room and also how little acoustic power is produced by the human voice.

7.10   SOUND TRANSMISSION

Transmission of sound from one room to another (or from an outside space into a room) is effected via several paths. The direct path through the common wall is usually responsible for the major part of the transmitted energy. The other routes for sound transmission are known as flanking paths. These occur as a result of a complex of transmissions of structural vibration and sound along lateral paths in the building. For the direct path, the mechanism of transmission is that the wall is set into motion on the source side and this motion, in turn, excites the air in the receiving room. In most circumstances, the frequency spectrum of the sound lies predominantly above the natural frequency of the fundamental mode of vibration of the wall. This mode of vibration is largely responsible for the transmission of sound.

When a simple mass–spring system is excited into vibration above its natural frequency, then the response is governed by the mass of the system (see chapter 1). Hence, in this case, sound transmission depends very largely on the mass of the wall. The greater the mass, the less will be the sound transmitted. This is the basis for the 'Mass Law' which states that for each doubling of wall mass the Transmission Loss (**TL**) increases by 6 dB. One aspect of this result is that increases in wall stiffness and damping are of little value in increasing Transmission Loss, contrary to a common misconception. A fuller mathematical statement of the mass law is obtained in the simple theory which follows.

This theory looks at harmonic excitation of the wall under the action of acoustic pressures:

$$p(t) = pe^{i\omega t}$$

It is assumed that the pressures act coherently over the wall surface (normal incidence). If it is only the mass of the wall which affects its vibration then it can be regarded as a 'limp partition'. That is to say, the bending stiffness of the wall is ignored because it does not significantly affect the motion. For each unit area (1 m^2) of the wall, of mass m, the net force F is

$$F = pe^{i\omega t}$$

This force causes acceleration of the wall, given by

$$m\ddot{u} = pe^{i\omega t} \tag{7.38}$$

The motion of the wall (u is displacement) will be

$$u = Ue^{i\omega t}$$
$$\dot{u} = i\omega Ue^{i\omega t}$$
$$\ddot{u} = (i\omega)^2 Ue^{i\omega t} \tag{7.39}$$

Substituting equation 7.39 in 7.38 gives

$$m(i\omega)^2 U = p \tag{7.40}$$

The peak particle velocity amplitude, v, is the magnitude of $\dot{u}$, which from equation 7.39 is

$$v = i\omega U \tag{7.41}$$

Combining equations 7.40 and 7.41, it follows that the specific acoustic impedance of the wall is

$$\text{impedance} = \frac{p}{v} = mi\omega \qquad (7.42)$$

This impedance is now used in an analysis, rather like the one at the beginning of this chapter, to find the Transmission Loss (**TL**).

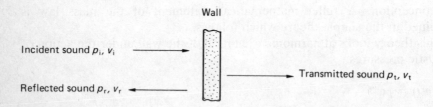

Figure 7.8

Referring to figure 7.8, where the symbols have the same meanings as described earlier for figure 7.1, the conditions at the wall are:

velocity

$$v_i + v_r = v_t \qquad (7.43)$$

force

$$p_i + p_r - p_t = (mi\omega)v_t \qquad (7.44)$$

From equations 1.14 and 1.15, the relationships between pressure and velocity are

$$p_i = \rho c v_i \qquad (7.45)$$

$$p_t = \rho c v_t$$

$$p_r = -\rho c v_r$$

By substitution of equation 7.45 in 7.44 and eliminating v_r using equation 7.43, the following result is obtained:

$$\frac{p_i}{p_t} = \left(1 + \frac{mi\omega}{2\rho c}\right) \qquad (7.46)$$

Using logarithmic units to describe the **TL** in dB:

$$\mathbf{TL} = 10\log_{10}\left|\frac{p_i}{p_t}\right|^2 = 10\log_{10}\left(1 + \left(\frac{m\omega}{2\rho c}\right)^2\right) \tag{7.47}$$

In almost all practical circumstances:

$$\frac{m\omega}{2\rho c} \gg 1$$

so equation 7.47 becomes:

$$\mathbf{TL} = 20\log_{10}\left(\frac{m\omega}{2\rho c}\right) \tag{7.48}$$

Using the known properties of air this can be restated:

$$\mathbf{TL} = 20\log_{10}m + 20\log_{10}f - 42 \tag{7.49}$$

A more accurate theory, taking into account all angles of incidence on the wall, indicates that the 42 dB figure should 48 dB.

This is the classic form of the Mass Law which shows that the **TL** increases by 6 dB each time the wall mass is doubled. Equally important is the 6 dB increase in **TL** with doubling of frequency. Low-frequency sounds transmit relatively easily; high-frequency sounds are greatly attenuated.

Example

A concrete partition wall is required to have a **TL** of 40 dB at a frequency of 100 Hz. Calculate the required thickness of the wall. The density of concrete is 2400 kg/m^3.

Solution

Assume that the wall thickness is t metres. The mass per unit area is then

$$m = 2400t$$

Using equation 7.49:

$$40 \text{ dB} = 20\log_{10}(2400t) + 20\log_{10}100 - 42$$

Hence

$$t = 0.0525 \text{ metres}$$

Although equation 7.49 makes quite accurate predictions of **TL** for many real walls and partitions, the situation is more complex than the simple assumptions of the theory imply and there are several points which need to be made. As has been already mentioned, the theory does not take account of random incidence of sound or of the multi-modal behaviour of the wall. Because of these, the Mass Law figures of 6 dB per doubling of mass and 6 dB per octave for frequency both decrease to about 5 dB. Figure 7.9 shows a theoretical line for **TL** against wall mass which takes account of these. Also shown are measured **TL**-values for various wall construction materials. It is seen that these real materials mostly fit the theoretical line but with some significant exceptions.

Double-leaf constructions, such as double glazing, are seen to perform significantly better than the simple Mass Law predicts. The reason for this

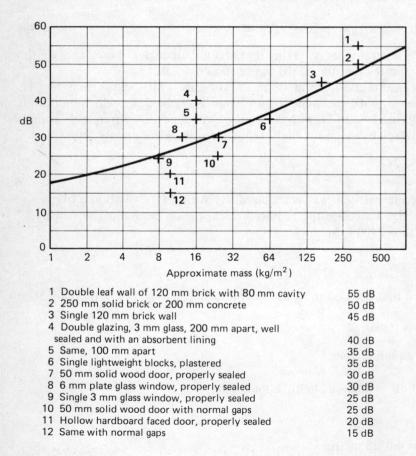

1 Double leaf wall of 120 mm brick with 80 mm cavity .... 55 dB
2 250 mm solid brick or 200 mm concrete .... 50 dB
3 Single 120 mm brick wall .... 45 dB
4 Double glazing, 3 mm glass, 200 mm apart, well
  sealed and with an absorbent lining .... 40 dB
5 Same, 100 mm apart .... 35 dB
6 Single lightweight blocks, plastered .... 35 dB
7 50 mm solid wood door, properly sealed .... 30 dB
8 6 mm plate glass window, properly sealed .... 30 dB
9 Single 3 mm glass window, properly sealed .... 25 dB
10 50 mm solid wood door with normal gaps .... 25 dB
11 Hollow hardboard faced door, properly sealed .... 20 dB
12 Same with normal gaps .... 15 dB

Figure 7.9 Transmission Loss values (av. 125 Hz – 4 kHz) for some typical building materials, compared with the modified mass law

can be understood by considering a simple example. Consider a construction consisting of two sheets of 3 mm glass. A single such sheet has a **TL** of 25 dB. Two stuck together to form a sheet of 6 mm glass have double the mass and so the **TL** should be 31 dB. Now consider the same two sheets, but separated by a gap. Each one has a **TL** of 25 dB so the total for the two should be $(25 + 25) = 50$ dB. Thus the gap is seen to be important; but how big should it be? The answer is that the gap should be at least a wavelength wide if it is to be fully effective. This is one reason why thermal double glazing, with its necessarily narrow gap, does not give the great acoustical advantage indicated by the simple calculation above. Compare the calculated values in this simple illustration with those given on figure 7.9.

The other exception to the rule indicated on figure 7.9 is that constructions which are not sealed perform less well. This is because the sound will pass through holes or gaps without attenuation. Even small holes can seriously degrade the performance. Take as an example a wall with a sealed **TL** of 45 dB. The energy is then attenuated by a factor of about 30000. If there is a hole which is only 1/30000 of the whole wall area, twice as much energy will be transmitted and the **TL** will be reduced to 42 dB. Thus it is seen that sealing, particularly around doors and windows, can be very important. Even a keyhole in a door can reduce the **TL**.

A further point of importance is the so-called coincidence effect. This occurs at a frequency for which the partition has a natural bending frequency *and* the wavelength in bending coincides with the acoustical trace wavelength (see figure 3.6 to understand this term). Under these conditions, sound may pass through the partition virtually without attenuation. This frequency is called the critical frequency (f_c in Hz) and for a wall of isotropic uniform material it may be calculated from the approximate formula

$$f_c = 58000/c_w t \quad \text{(Hz)} \tag{7.50}$$

where c_w is the wavespeed in the wall material in m/s (see table 1.1) and t is the wall thickness in metres. Thus, for a single brick wall:

$$c_w = \sqrt{(E/\rho)} \approx 3000 \text{ m/s}$$
$$t = 0.1 \text{ m}$$

and hence

$$f_c = 193 \text{ Hz}$$

Figure 7.10 shows a typical measured curve of **TL** against frequency for a partition. The effect of a critical frequency at about 1 kHz is shown. It can be controlled by the use of vibration damping though this is not always easy

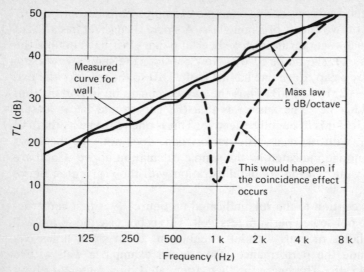

Figure 7.10   Typical wall **TL** *as a function of frequency*

for many types and materials of construction. The reader interested in more detail of this effect should consult reference [1] where a wealth of information is given on the subject.

The **TL** is a property of the wall or partition alone. In practice, the room or rooms on either side of the wall will be more or less reverberant. Thus, if measurements are made, *in situ*, of the Noise Reduction (**NR**) obtained across the wall, it will usually be less than the **TL**. Predictions for the receiving room may be made from the formula:

$$NR = TL - 10\log_{10}\left(\frac{S}{A}\right) + 0.3 \qquad (7.51)$$

where *S* is the area of the wall and *A* is the total absorption in the receiving room. The **NR**-value may be further reduced by flanking path transmission. This kind of transmission can be reduced at the design stage by inserting discontinuities and vibration absorbing layers in the primary structure of the building.

A useful set of approximate empirical rules for adjacent offices has been communicated to the authors as follows:

Offices abutting end to end: **NR = TL + 3 dB**
        (shorter walls)

Offices abutting side to side: **NR = TL**
        (longer walls)

Offices one above the other: **NR = TL − 3 dB**

These can be derived from equation 7.42, if suitable assumptions are made, and have been found to be widely verified by experience. However, there must always be the proviso that final values can be derived properly, given the necessary data for the rooms in question.

The analysis of sound transmission is important not only for sounds transmitted into rooms but also in the design of machine enclosures, where exactly the same principles apply. The design of enclosures for machines often requires entry points for services and forced cooling systems. These points may provide large air gaps in the enclosure which will degrade its acoustic performance, for the reasons given above. Consequently they require special attention at the design stage.

REFERENCES

[1] D. A. Bies and C. H. Hansen, *Engineering Noise Control*, Unwin Hyman, (1988).
[2] B. J. Smith, R. J. Peters and S. Owen, *Acoustics and Noise Control*, Longman (1982).
[3] *British Standard Code of Practice CP3, Code of basic data for the design of buildings; Chapter III, Sound insulation and noise reduction*, British Standards Institution (1972).

8. Noise control

No other knight in all the land
Could do the things which he could do
Not only did he understand
The way to polish swords, but knew
What remedy a knight should seek
Whose armour had begun to squeak.
A. A. Milne, *Now We Are Six*

8.1 INTRODUCTION

In common parlance the word 'noise' means sounds which are annoying and undesirable. Very often these undesirable sounds are also noises in the technical sense of having a broad frequency spectrum. The function of the professional noise control engineer is to minimise this noise. The preceding chapters have outlined the material which is necessary to an understanding of the physical generation, transmission and measurement of noise. This chapter shows how this material can be put to good use by the noise control engineer in solving practical noise problems in all their variety.

Two fundamental kinds of problem may be encountered. In the first the engineer is required to minimise noise in a *proposed* situation by means of *design*. This involves planning for noise and is sometimes rather vaguely described as 'noise management'. In practice this requires a questioning approach to potential problems (for example, can this process be done another way?) together with an implementation of what is considered to be 'good practice'. In the design of a commercial product it may also require a consideration of cost, of competition with other similar products and, in some cases, of meeting an industry standard for noise emission. The second, and much more common, type of problem is concerned with a situation which already exists. The perceived noise is considered, by some criterion, to be so great that something must be done about it. This is the 'retrofit' problem. This chapter is mostly about solving retrofit noise problems, although many of the principles will constitute 'good practice' when applied to design.

8.2 NOISE-CONSCIOUS DESIGN

The special factors involved in planning for a noisy situation are now briefly considered. These are listed below in the context of planning the design of some complex machinery. Similar considerations are also relevant to noise problems outside engineering, such as in road traffic planning or the siting of a pop festival. The factors are presented in the form of a list of questions:

- *Design*. Is full advantage being taken of existing knowledge of noise-reducing measures such as vibration isolation, enclosures, mufflers, barriers, etc.?
- *Selection*. If an outside manufacturer is supplying crucial parts, has (a) a noise and vibration specification been evaluated for the component, (b) an alternative and quieter method for achieving the same purpose been considered?
- *Siting*. Are there (a) better alternative layouts for the components in the machine, (b) better alternative positions for the machine relative to the people who are going to be affected, (c) better alternative structural or service-line connections (direct connections to large surfaces such as walls and floors are acoustically undesirable, see chapter 3)?
- *Operation*. Are there possible variations in operating procedures or times which can mitigate potential nuisance (for example, can the machine be run under automatic control with no nearby operator; can it be run intermittently)?

Once the machinery is installed there are also useful actions which can be taken to minimise any subsequent noise problems. The noise will inevitably increase as time goes by, owing to wear; so it is important to have a proper maintenance programme with a strong sense of noise-consciousness. Many problems can be avoided by encouraging operatives to understand the noise-making process and by encouraging them to make use of any facilities provided such as ear protectors and quiet rooms.

8.3 EXISTING NOISE PROBLEMS

The remainder of this chapter will consider in detail how the retrofit type of problem may be resolved. First, it is necessary to assess the magnitude of the problem. Noise measurements must therefore be made and compared with an appropriate criterion. There are many different criteria, often couched in different noise units, to cope with the wide variety of problems which may arise. The criteria may also vary from one country or culture to another. In many cases, where an understanding of human response to the noise is incomplete (such as, for example, with aircraft noise) the criteria are in a state of development. An account of criteria relevant to many noise problems is given in chapter 2. These criteria have been stated in rather general terms, though some benchmark figures have been given by reference to specific standards. This has been done so that the reader may have some idea of appropriate critical levels in various situations.

Having estimated the magnitude of the problem, it is then necessary to decide what to do about it. For this purpose it is essential to have a clear physical picture of the noise-generating system. The next section of this chapter describes the structure of noise-generating systems and builds on the foundation of first principles described in previous chapters. The methods available for noise reduction within the system are listed. Once the physical structure of the noise system has been determined, dB-values must be attributed to each part of the system. This is the skilled process of *diagnosis*. Without an adequate diagnosis, it is not possible to treat the problem effectively. There are a number of tools and methods which are available to the noise control engineer in order to make an accurate diagnosis. One of these strategies, called the series–parallel method, is described here. This shows how diagnosis may lead to an optimum solution to the noise problem. This solution not only makes effective use of noise control technology to design the best means for reducing the noise but also considers the cost-effectiveness of the treatment and whether or not noise targets can be met. A note of caution: in making changes to a system in order to achieve noise control there are also secondary considerations. The changes can lead to a deterioration in safety, chemical compatibilities, cooling (for machines), life to failure, and the risk of increased noise and vibration elsewhere.

8.4 ACOUSTIC SYSTEMS

A system which produces an undesirable noise must be correctly analysed into its component parts if suitable actions are to be taken to alleviate the noise. All noise systems may be subdivided into three parts: see figure 8.1. For many noise systems, particularly machines, the SOURCE element in this diagram is further reducible to three parts, as shown in figure 8.2

It is only where the source causes air disturbances directly, without mechanical moving parts (for example, with gas jets), that this pattern is not followed. Thus for most sources a noise-producing system is regarded as a five-part system. Interference with any part of this system will cause change and hence offers the possibility of reduction of the noise. However, it is usually best to interfere with elements at the left-hand (source) end of

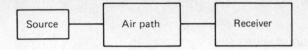

Figure 8.1 *The structure of acoustic systems*

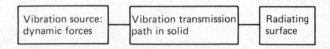

Figure 8.2 *The structure of an acoustic source*

the chain first because there are often parallel paths of transmission. Thus, vibration control is often as important as noise control, and this explains why the two subjects are so intimately linked.

Table 8.1 shows some typical examples of problems which arise in machines as noise sources. It also shows the type of action, based on the fundamental principles described earlier, which can be taken to resolve the problem. Table 8.2 regards the source as a single entity and shows the type of palliative which may be considered in reducing the noise from the system as a whole.

The general structure of acoustic systems should now be clear. Tables 8.1 and 8.2 have listed some of the actions which can be taken to reduce the noise from individual components of the system. How then can a specific problem be analysed? The steps in the process are as follows:

(1) Sketch the whole noise-producing system from source to receiver.
(2) Determine the measures which can be taken with existing elements in the system to reduce noise. With this information and a full set of diagnostic measurements, the overall noise reductions can be determined (see the next section). Determine the cost of these measures.
(3) Calculate whether the possible modifications to the system, either individually or together:
    (a) meet the noise specification;
    (b) fall within the financial limit.
(4) Pose the question, whether it is cheaper to scrap the system in favour of a quieter one.

The following section goes into the detail of one method for implementing these steps. By this means the structure of the system is analysed, a detailed diagnosis of the problem is obtained and the cost-effectiveness of solutions is assessed.

Table 8.1 Machines as noise sources.

Structure	Examples	Palliatives
Source of dynamic force	Out of balance (inertia forces)	Static and dynamic balancing
	Magnetic	Design of pole-pieces and other components
	Combustion	Minimise pressures and their rate of variation
	Sliding and rolling contacts	Smooth surfaces and low contact stresses
	Aerodynamic	Encourage smooth flow: avoid turbulence and instabilities
	Impact	Minimise peak pressures: maximise period of contact
	—	Use machines working on different principles
Vibration transmission path	Structural connections; internal and external	Use isolation (see chapter 1). Use impedance mismatch. Control mass, stiffness or damping (see chapter 1). Use vibration absorbers
Radiating surfaces	Casings, foundations and other attached surfaces such as walls or hoppers	Minimise area (see chapter 3). Maximise isolation. Use damping material on thin panels. Use isolated covers

Table 8.2

Structure	Examples	Palliatives
Acoustic source	Fans and air conditioning systems (ducts)	Run as slowly as possible. Keep moving blades away from static components such as struts. Use absorbing materials, Helmholtz resonators and silencers (see chapter 7). Active noise control
	Exhausts and other intermittent gas flows	Use reactive and/or dissipative mufflers (see chapter 7)
	Burners and jets	Minimise flow velocities and the shear gradient at the jet edges
	Machines	See table 8.1
Acoustic transmission path	—	Use machine enclosures (see chapter 7)
	—	Use walls or barriers (see chapter 3)
	—	Consider use of acoustic absorbing material for enclosures and in rooms
	—	In the open, align the source to make use of directionality
	—	Increase distance from the source
Receiver	—	Use personal ear defenders: shorter working hours

8.5   ANALYSIS USING THE SERIES–PARALLEL METHOD

"What would life be without arithmetic,
but a scene of horrors!"
Sydney Smith

It is often not at all clear to the noise control engineer which component of a noisy machine is the principal source. Even when it is clear, the amount of attenuation which is to be obtained by silencing that component is difficult to assess because this requires a knowledge of the 'remainder' noise which the machine is making. There are two alternatives which may be adopted:

(1) to control the greatest source of noise and hope that the attenuation obtained is sufficient, or,
(2) to embark on tests designed to discover the 'remainder' noise by measuring the noise from the machine when the noisiest component is disconnected or effectively muffled.

The latter is clearly a more satisfactory diagnostic procedure where time and the structure of the machine permit. In this section an extended version of the procedure is illustrated by means of a case study, in which a complete analysis is made of a noise source by means of selective disconnection or muffling of the various parts. Of course, it may prove impossible to distinguish one source from another by such means; but then if that is so, it follows that it is impossible to attenuate one without the other. In such circumstances it may be sensible to regard the two sources as one.

By these means not only may the remainder noise be determined, but also the effect of various alternative noise-reducing measures may be computed. Above all, the cost of each of the possible noise control measures may be used to study their cost-effectiveness. If the aim is to satisfy a particular set of noise regulations for a minimum cost, as is often the case, then this may easily be achieved. The scheme is particularly well adapted to the solving of noise problems for engineering plant which is built up from a series of discrete machines because it is often relatively easy to disconnect individual components of the plant for test measurements or for replacement with a less noisy component. In the method described here, the plant is analysed into what are termed 'series and parallel noise sources' (see reference [1]).

The logic of the scheme follows quite simply from the analysis in the previous section. A single point of dynamic stress inside a machine, such as a rumbling bearing, a pole of an electric motor or the explosion in a cylinder head, will give rise to stress waves which radiate throughout the structure of the machine and to points and surfaces connected to it. Several surfaces may thus be significantly responsible for the radiation of noise from this force source. Such surfaces are a *series* set of acoustic sources because they are jointly responsible for the sound radiation originating from a single force source. These surfaces are termed 'acoustic sources', as distinct from 'force sources', and are shown on the diagrams as small circles with an identifying code within. Another force source, a gearbox for example, will be responsible for significant noise radiation from a different set of surfaces (some of which may be the same as before); these constitute a different set of series acoustic sources. Each set of series acoustic sources thus depends on one of a *parallel* set of force sources. The arrangement described may be appreciated more readily by reference to figure 8.3 which shows how the noise radiation from a simple gearbox may be displayed, the two parallel paths corresponding to noise arising from rumbling bearings and from meshing in the gear box. Meshing noise radiating from the subframe surface is assumed to be insignificant.

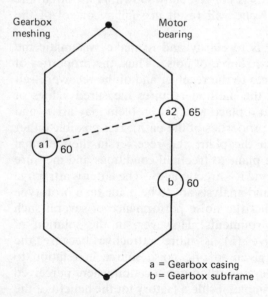

Figure 8.3

Each parallel path includes the various acoustic sources associated with that path and a figure is shown giving the noise figure associated with that acoustic source. Thus the total noise as shown is

(60 + 60 + 65) dBA = 67.1 dBA   (use the nomogram, figure 1.6)

It is now very easy to determine the effect of any noise-reducing measure. For example, if the gearbox were to be isolated from the sub-frame to give an attentuation of 30 dB, the total would then be $(60 + 30 + 65)$ dBA = 66.2 dBA. It can be seen that this would not be an attractive solution because the total noise is only reduced by 0.9 dBA.

Generally speaking, control of any radiating surface, such as may be achieved by enclosure, will only reduce the noise radiation at that one point in the diagram. Control of dynamic loads at source, such as replacement of the rumbling bearing, will affect all of the series noise sources associated with it. The diagram also quickly indicates what residual noise there will be when a particular noise-control measure is taken. This example shows that the full benefit of a particular noise-control measure is not usually reflected in the reduction of total noise. Another point of importance is that the background ambient noise level should be included in the diagram as a parallel path. One feature of this simple example is that acoustic sources (radiating surfaces) may appear in two or more of the parallel paths so that control of this surface may have a double or triple effect. This link between sources is conveniently shown by a dotted line which indicates that control of one will result in similar control of the other.

The purpose of the method is to clarify and to make systematic the nature of the machine alone as a source of noise. Thus, the properties of the path from the acoustic sources to the receiver and of the receiver itself are not considered. However, the method requires measured values of noise and they are best taken at a place where a problem may arise (and therefore implicitly include the properties of the path). It is possible either (a) to measure the sound with the plant and receiver in their normal positions, or (b) to remove the plant to freefield conditions and measure the total sound power radiated and its directionality. The latter is attractive where the noise measurement and analysis are to be made on a prototype and subsequently used to predict the noise performance of several such machines in a variety of environments. However, in the solution of particular problems, alternative (a) is more attractive because the appropriate emphasis will be given to one noise source in relation to others. Of course, there may be more than one location where perceived noise is to be controlled, for example, inside a factory for the benefit of the workers, and outside for the benefit of local residents. In such a case two analyses of the same noise producing system are necessary.

8.6   CASE STUDY OF THE NOISE FROM A WOOD FIBRE MILL

A more complex example illustrating a real noise problem is now given in which a noise source analysis has been made of a machine which grinds wood off-cuts into small fibrous particles for subsequent use in the manufacture of fibre board. Two problems arose with this machine: first, the noise level at the operator's position (92 dBA) was in excess of the legally permitted value (90 dBA); secondly, the noise from the machine was causing distress to local residents. For reasons of clarity and limited space, the second problem will not be considered here. An amount of money not exceeding £750 was allocated to relieve the problem. The structure of the machine together with a list of acoustically radiating surfaces is shown in figure 8.4.

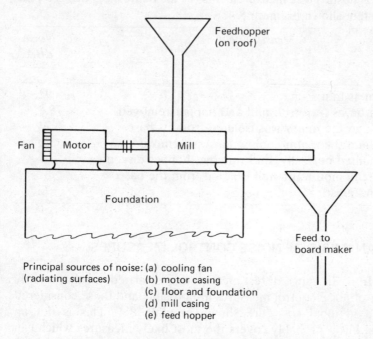

Figure 8.4   Schematic layout of the wood fibre mill described in the case study

Diagnostic measurements were made under various conditions to determine the noise levels associated with each radiating surface at the operator's position. These are shown in table 8.3. The figures given were then used to construct the series–parallel diagram shown in figure 8.5. To deduce the values shown in figure 8.5 from the figures shown in table 8.3 requires the solution of a set of simultaneous equations. This process makes some assumptions. First, it has been assumed that the electric motor run freely will produce as much noise as it would do when driving the mill. This will clearly not be the case, but the magnitude of the discrepancy is uncertain. Another assumption made in this type of analysis is that interaction effects are negligible: for example, that connection or disconnection of the mill to the motor will have no effect on the way that motor noise is radiated from the motor casing or from the foundation. One way of minimising errors which arise from these assumptions is to carry out redundant diagnostic tests. It is then possible to analyse the results to minimise the errors, but the mathematical procedures required are complex. In this problem it will be noticed that seven tests have been carried out to find the value of various combinations of the seven acoustic sources.

Table 8.3 Diagnostic noise measurements (to the nearest 0.5 dBA) on the system shown in figure 8.4.

Measurement	Noise dBA
1. System run as first seen	92
2. Motor run on its own with mill and hopper removed	75.5
3. Ditto (2), and the motor was isolated from its base	74
4. Ditto (3), and the motor cooling fan was removed	70
5. Motor mounted normally, mill running, hopper disconnected	91
6. Ditto (5), and motor and mill isolated from the base	87
7. Background noise	72

8.7 COST ANALYSIS OF NOISE CONTROL MEASURES

It is obvious from the range of test measurements carried out that a fairly large number of noise-control measures are possible and those considered together with notional costs are shown in table 8.4. This is not an exhaustive list but it probably covers the most likely measures which can easily and reasonably cheaply be adopted.

Table 8.4 Four alternative noise control measures considered for the wood mill problem together with costs.

1. Isolation of hopper from mill with flexible section of tube: 30 dBA attenuation of hopper noise	£100
2. Isolation of motor and mill from foundation: 20 dBA attenuation of floor noise	£200
3. Ventilated acoustic enclosure for mill: 20 dBA attenuation of mill casing noise	£500
4. Ventilated acoustic enclosure for mill motor combination: 20 dBA attentuation of mill and motor casing noise	£600

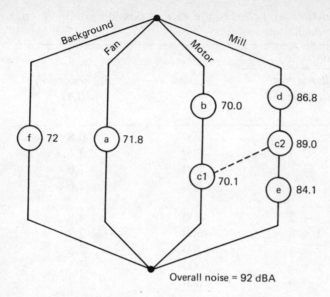

Figure 8.5 Series–parallel source analysis at the operator's position. See figure 8.4 for the letter-coding of acoustic sources

No costing of the labour involved in diagnosis has been included in these figures. All costs shown in this study are notional and based on 1989 values.

These measures can be adopted either singly or in any combination, so that the four possibilities would generally give rise to fifteen alternatives (this is $(2^n - 1)$, where $n = 4$). In this case the alternative measures 3 and 4 in table 8.4 cannot both be applied together so that there are only eleven valid alternatives. It can readily be appreciated that an analysis of all the alternatives when there are more than about five possible noise control measures becomes a cumbersome task best performed by computer. It is often obvious that some measure is not going to provide significant additional noise control. If the problem is being solved by hand, then it is usually best to adopt a pragmatic approach and concentrate on the largest source of noise first, working on from there to consider suitable combinations. In any case, the effects of the insulation measures suggested are very easily determined by reference to figure 8.5 and it is particularly here that the use of the series–parallel diagram is valuable.

Table 8.5 Noise reductions and costs of the various possible combinations of noise-reducing measures

Combination		Cost	Noise reduction
Letter code	Alternatives as in table 8.4	(£)	(dBA)
a	1	100	0.8
b	2	200	3.1
c	3	500	1.6
d	4	600	1.6
e	1 + 2	300	4.8
f	1 + 3	600	2.7
g	1 + 4	700	2.8
h	2 + 3	700	7.1
i	2 + 4	800	7.2
j	1 + 2 +3	800	14.7
k	1 + 2+ 4	800	15.6

The best way of presenting the noise control/cost information thus obtained is by means of a graph of noise reduction against cost for the various alternatives. The eleven alternatives considered in this study, together with corresponding costs and noise-reduction figures deduced from table 8.4 and figure 8.5, are shown in table 8.5. Figure 8.6 presents the same information graphically and points on it are letter-coded to correspond with table 8.5.

The application of noise-control criteria is then relatively straightforward: if there is a specific noise goal to be achieved then the attenuation required is known, as shown in figure 8.6. The cheapest option above this line can then be chosen. If maximum control is to be achieved within a budget then a vertical line will indicate the budget limit and the best combination of measures can be chosen (and may prove considerably cheaper than the allowed sum). In this problem it is clear from figure 8.6 that alternative (b) best meets the requirement for a sum of £200. One of the features of noise control work, which shows up on figure 8.6, is the law of diminishing returns, in terms of dBA/£, as more and more noise-control measures are used. Inevitably this law must apply because of the maximum possible noise reduction set by the background noise.

The use of the series–parallel diagram, though not without its difficulties and errors, is a useful approach to diagnosing noise problems and solving

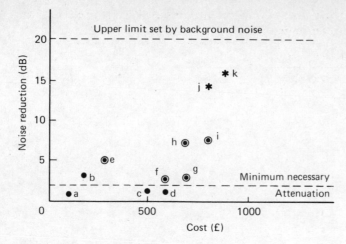

Figure 8.6 Noise reduction/cost figures for the options in table 8.5. See table 8.5 for letter-coding on this figure

them effectively. It also leads to a clear understanding of the detailed nature of noise problems with machines.

REFERENCE

[1] A. J. Pretlove, Series and parallel noise sources, *The Chartered Mechanical Engineer*, November 1976, 87–90.

Index